# Single Camera Stereo Sound

# Single Camera Stereo Sound

John Ratcliff
and
Neil Papworth

FOCAL PRESS

Focal Press
An imprint of Butterworth-Heinemann Ltd
Linacre House, Jordan Hill, Oxford OX2 8DP

PART OF REED INTERNATIONAL BOOKS

OXFORD LONDON BOSTON
MUNICH NEW DELHI SINGAPORE SYDNEY
TOKYO TORONTO WELLINGTON

First published 1992



**British Library Cataloguing in Publication Data**
Ratcliff, John
Single camera stereo sound.
I. Title II. Papworth, Neil
621.3893

ISBN 0 2405 1307 X

**Library of Congress Cataloguing in Publication Data**
Ratcliff, J. A. (John Ashworth)
Single camera stereo sound/John Ratcliff and Neil Papworth.
p. cm.
ISBN 0 2405 1307 X
1. Sound – Recording and reproducing. 2. Stereophonic sound systems. I. Papworth, Neil. II. Title.
TK7881.8.R36 1992
621.389'3–dc20 91–45194
CIP

Composition by Genesis Typesetting, Laser Quay, Rochester, Kent
Printed and bound in Great Britain by Redwood Press Limited, Melksham, Wiltshire

# Contents

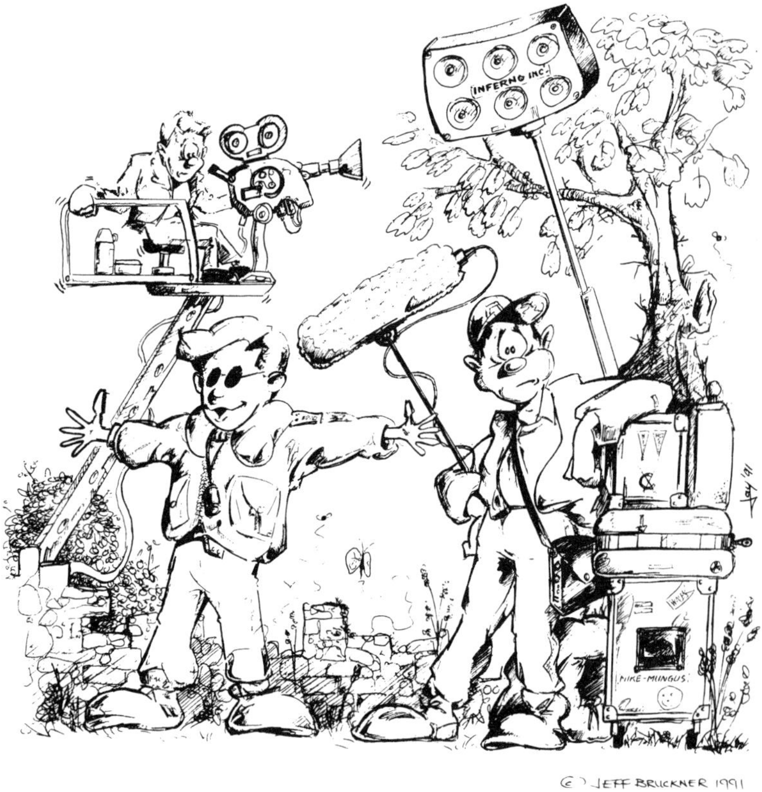


"I want it in stereo!"

# Introduction

The recording of sound on location is a delightful mixture of technical competence, artistry and compromise. There is never enough time and rarely enough people. The ability to work quickly and in impossible circumstances is paramount. You must, however, deliver the goods.

Despite some comments to the contrary, stereo recordings generally take more time than mono recordings. To say stereo can be 'slotted in' is a cause argued by people more interested in the challenge than in getting it right. That is not to say that in many situations stereo operations cannot be totally invisible. But overall stereo does, obviously, take more equipment, more time, and most importantly, more thought.

Stereo's greatest promise is to give us enhanced realism by allowing us to reproduce a much more accurate acoustic. We live in a stereo world and stereo does give us the opportunity to increase viewer involvement. But this gain in atmosphere causes us some of our greatest problems both on the road and in post-production. As we begin to plan for stereo most people immediately think of the problems associated with positional change. These are large, but obvious, even if the solutions are not. However there are other problems to consider that are more subtle and yet just as damaging to the overall effect.

We hope that by dispelling the myths that have come to surround the subject and by giving you a number of simple guidelines we can help to make stereo television as rewarding for you as it is for the viewer.

This is not a book of rules. Experienced recordists know that each situation needs to be worked in its own unique way; there are simply too many factors for every possibility to be discussed in detail. Anyone who tells you there is a 'right way' to work a scene has probably never done it. There is only one rule in our book. Does the sound add to the illusion? If it does we are winning!

# 1

# Stereo theory

## Directional hearing

The human ear/brain combination uses a number of clues to locate a sound in space. Chief among these are differences in arrival time, level and tonal quality of the sound from the source when it arrives at our two ears.

A sound directly in front of the listener will arrive at the two ears at the same time. So powerful is this timing element in determining the position of a sound that it will tend to override the other two clues even if one of the ears is partially blocked, with resultant differences both in level and tonal quality.

Sound travels relatively slowly, certainly compared to light, at approximately 340 metres/second, so a sound arriving from off-centre will arrive at one ear before the other (Figure 1.1). So sensitive is human hearing to this time element that differences in arrival time of as little as 10 microseconds can be detected. This represents a sound source approximately just one degree off-centre.

In addition to timing differences, high frequency sounds from sources considerably round to the side of the head will be attenuated before reaching the far ear, as their wavelength will be shorter than the dimensions of the head, which will act as an obstacle (Figure 1.2). In fact the directivity of the human ear/brain combination is greatest at about 700 Hz, the frequency whose half-wavelength is comparable to the spacing between our ears.

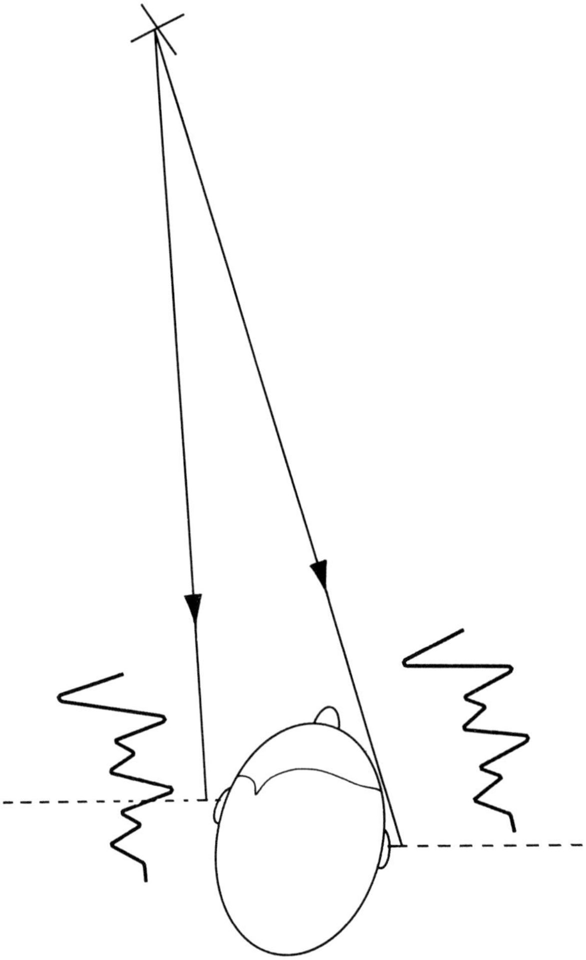

**Figure 1.1** Sound arriving from a source left of centre will arrive at the left ear first, creating a 'time of arrival' difference between left and right ears.

As regards front/back and up/down location of sources, the situation is not so clear cut, but it is generally considered that the Doppler effect (the effect that causes an apparent pitch change in sounds that approach and recede), brought about by small movements of the head, enables front/back differentiation, possibly aided by interference effects brought about by the

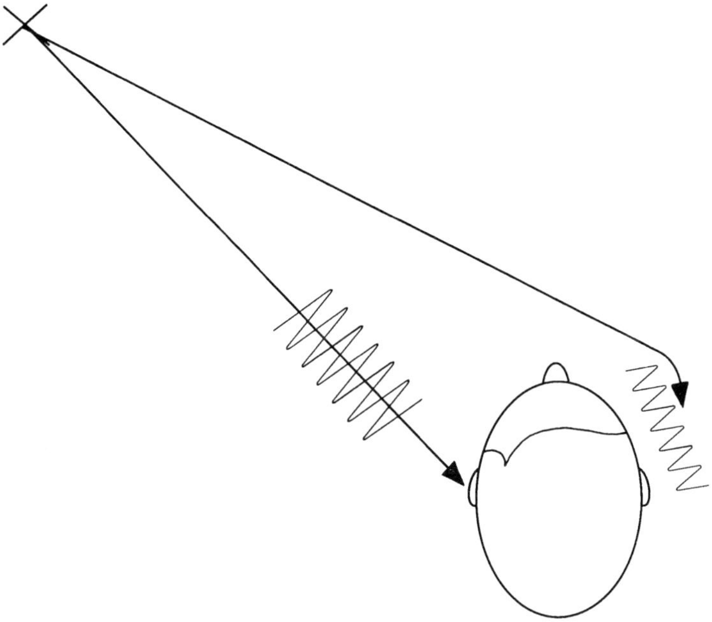

**Figure 1.2** For wavelengths shorter than the diameter of the head the sound to the further ear from a source substantially off-centre will be attenuated.

convolutions on the external ear flap (the pinna). The effect will be reversed when comparing front and rear sound sources. Up/down differentiation is generally thought to be due to learned experience (we do not expect to hear footsteps coming from above, or aircraft from below), and to differences in the nature of the reflected *vis-à-vis* the direct sound (the sky does not reflect sounds coming from below, the ground will reflect sounds coming from above).

In practice it is found in some situations that contradictions in timing and amplitude information reaching the ears can cause sounds to be mislocated, an effect which has to be taken into account when considering sound sources placed to the sides of stereo microphones. For example, the well-known Haas effect will tend to cause the ear to locate a common sound from two sources towards whichever source provides the sound leading in time, even though the other source may provide a sound higher in amplitude (an effect commonly exploited in sound reinforcement systems). Of particular importance to the stereo sound recordist is that sounds arriving at a stereo pair of microphones with more than 90° of phase shift may well be reproduced outside of the baseline of the reproducing loudspeakers.

## Stereo reproduction

Although it is differences in time of arrival between the two ears of a single source (or a number of single sources in real life) that give the major clues in locating a sound in space, experiments by Alan Blumlein in the early 1930s showed that two speakers fed with a common signal, but at different levels, can create the effect of timing differences between the ears, and thus the illusion of a single sound source, whose position between the speakers depends on the level differences between them. The speakers should ideally be placed in front of the listener, spaced apart so that they form an equilateral triangle with the listener. Figure 1.3 shows how these two signals combine to create the effect of a single signal with timing differences between the ears.

It is worth remembering that this stereo effect does rely on what we might call these 'crossed' signals (left loudspeaker into right ear and *vice versa*), and that the sound recordist on location, particularly when working with a portable single camera unit, is almost certainly going to be listening on headphones, where these crossed signals will not be present. A number of attempts have been made to develop headphones that will provide these crossed signals, but none to date has been particularly successful.

Another fact worth remembering is that the stereo effect is easily destroyed by, for example, the listener moving off-centre, when not only does the created image appear to shift, but the extra timing differences brought about by being closer to one speaker than the other cause the image to be less well defined. Turning the head also causes the image to move, for similar reasons.

Because the image is easily destroyed by unwanted differences in level and timing, it is important that at all stages of production, post-production and transmission every care should be taken to ensure that unwanted level, phase and timing errors are avoided.

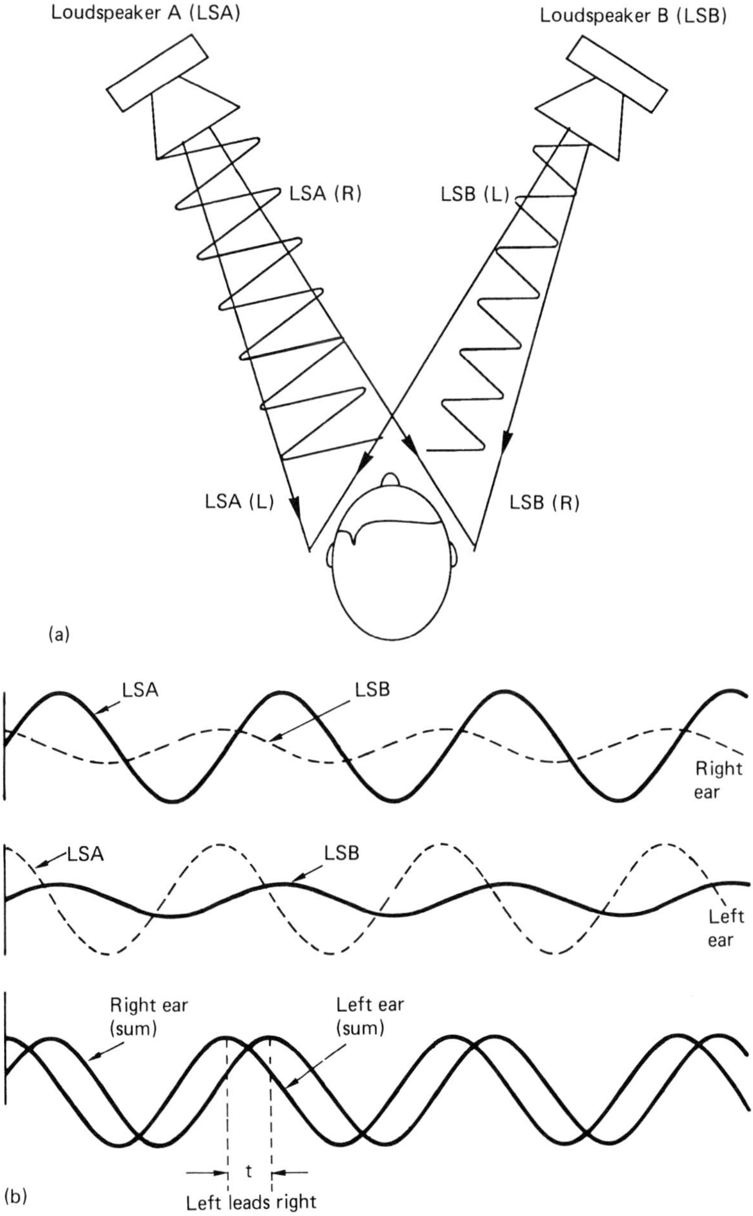


**Figure 1.3** LSA and LSB are reproducing signals identical in every respect other than amplitude. The signals from each speaker reach each ear (a). The amplitude and time-of-arrival differences combine to produce an effect akin to that from a single off-centre source (b).

### *AB stereo signals*

In broadcasting, stereo signals are often processed as separate LEFT and RIGHT, passing through identical, and usually adjacent, channels on sound mixing desks, distribution amplifiers, jackfield sockets etc. From the foregoing arguments it can be appreciated that such component parts of the system should be matched very closely to each other if stereo image position and mono compatibility are to be maintained.

The convention in broadcasting is for these left and right signals to be referred to as 'A' for left and 'B' for right. The reason for this designation as A and B is to avoid confusion when the two channels of a stereo system are not carrying a stereo signal (there is provision in the system for the two channels to carry separate languages, or for one channel to carry digital data).

### *XY–AB confusion*

Some authors use the XY abbreviation to denote left and right microphones. Others only call crossed cardioids XY. Mixing is then referred to as AB. In this book we will use A for left and B for right, throughout.

Professional convention is for the A signal to be indicated by RED (on cabling, faders, meters etc) and for the B signal to be coded GREEN. This can be confusing when domestic equipment uses red as the code for right. Probably the best way to be sure of the correct professional convention is to remember that the colour coding for left and right, or AB, signals is the same as for navigation lights.

### *MS stereo signals*

It is perfectly possible to view the stereo signal not as one having left and right components, but as one having a forward-facing component which will be strongest for a source placed in the centre of the sound stage, and a side component which will be strongest for a source placed to the side of the sound stage.

The forward component will give the signal that will be received by a monophonic listener, and is regarded as being the SUM of the left and right signals. It is generally known as the 'M' (standing for MAIN) signal and its colour code is WHITE, the same colour as existing mono system meter needles.

The 'S' or SIDE component is that which gives the directional information. This side signal is regarded as being the DIFFERENCE between the left and right channels, and will thus be at a minimum for a sound source placed in the centre of the stereo sound stage. Its colour code is YELLOW, possibly a poor choice where television is concerned since white also looks yellow under the dim lighting conditions which often prevail in television sound control rooms. However, one meter manufacturer supplies an MS meter with a white M needle and a black-with-red-tip S needle which is a lot easier to read in poor light. The side signal will only be picked up by listeners in stereo. Thus it is important that primary sound sources are not placed too far to one side, where the side signal will only be enhanced at the expense of the M signal, to the detriment of listeners in mono. Figure 1.4 illustrates the differences between AB and MS techniques.

Since the main and side components of the complete MS stereo system can be regarded as being respectively the sum and the difference between the left and the right components, the term 'sum and difference', when referring to MS microphones or matrices, has become common parlance.

## *Mono/stereo level compatibility*

It can be argued that if a centre-placed signal (i.e. one that has equal amplitude in left and right channels) is derived from a straight sum of the A and B signals, then, if these A and B signals are equal both in amplitude and phase, the resulting increase in voltage level (a doubling or 6 dB rise) may cause over-modulation of the mono channel. In fact the A and B channels of a stereo pair are rarely identical in these respects and the accepted convention is to apply a correcting factor to compensate, though there is no agreement between different organizations as to what this compensating factor should be. 0 dB, 3 dB and 6 dB are all in use. In practice, it has been found

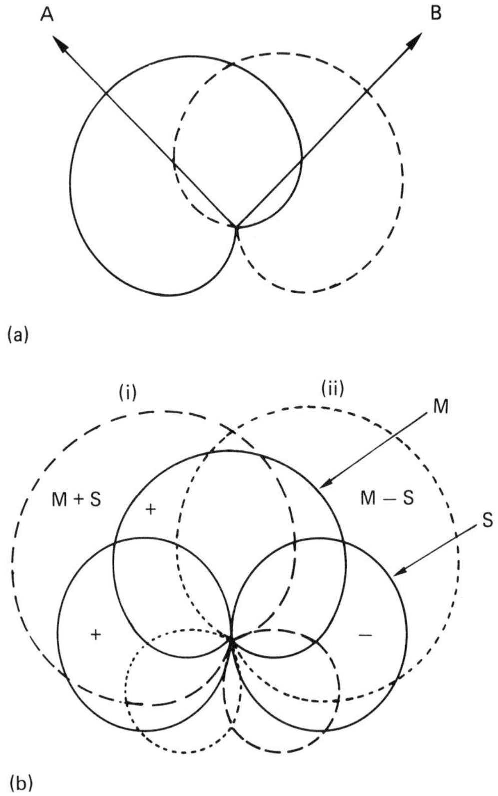


**Figure 1.4** Angled cardioid microphones, placed coincidently, will recreate amplitude differences for off-centre sound sources (a). If a forward-facing cardioid (M) is combined with a sideways-facing figure-of-eight (S) their sum (i) will produce a leftwards-facing hypercardioid, and their difference (ii) a rightwards-facing hypercardioid (b).

that with a number of common microphone techniques, particularly those in use for location recording, very little of the sound energy is coherent (i.e. identical in level and phase between the left and right channels), so no correcting factor need be applied.

It has been argued that because a listener in stereo hears the sum of the POWER from the A and B channels (doubling of power from a centre-placed source gives a 3 dB increase), whereas the mono listener hears the sum of the VOLTAGE (6 dB for a centre-placed source), there will be a difference in the centre/side balance presented to listeners in stereo *vis-à-vis* those with mono (Figure 1.5). Since neither type of listener is likely to be switching between mono and stereo this is hardly likely to matter. Indeed, it can be thought of as an advantage in that images placed towards the sides will appear lower in the mix for the mono listener than for the person with stereo, which is probably what is wanted – particularly if these side images are there to enhance 'reality', e.g. effects.

## *Dynamic use of the S signal*

The S signal (A–B), again with a 0 dB, 3 dB or 6 dB correction factor, will provide width to the stereo signal. Without the S signal there can be no directional information present. As the S signal is increased in level, so the stereo information increases. Thus the S signal can be used as a width control. The facility to vary the width of a stereo sound picture is very important in broadcasting, particularly in television where not only does there have to be continuity between successive scenes, but where the sound image also has to complement the visual one.

Although MS techniques are not widely found in the music recording industry, they are common in broadcasting as not only can image width be easily varied at any stage prior to transmission, but also where mono compatibility is of great importance, it can be easily checked.

Many stereo microphones used in broadcasting, particularly those used on location recording, provide both AB and MS outputs (switchable between the two) and some control over the relative sensitivities between the M and S elements, with an indication of the acceptance angle thus provided. Many location sound recordists with experience in working with stereo regard this 'width' control as an operational one, as opposed to a preset, to dynamically control the width during a take. Figure 1.6 illustrates such a microphone together with its control box.

## *Conversion between AB and MS signals*

If either A and B or M and S signals are needed at some point in the production process it may be necessary to provide them. For

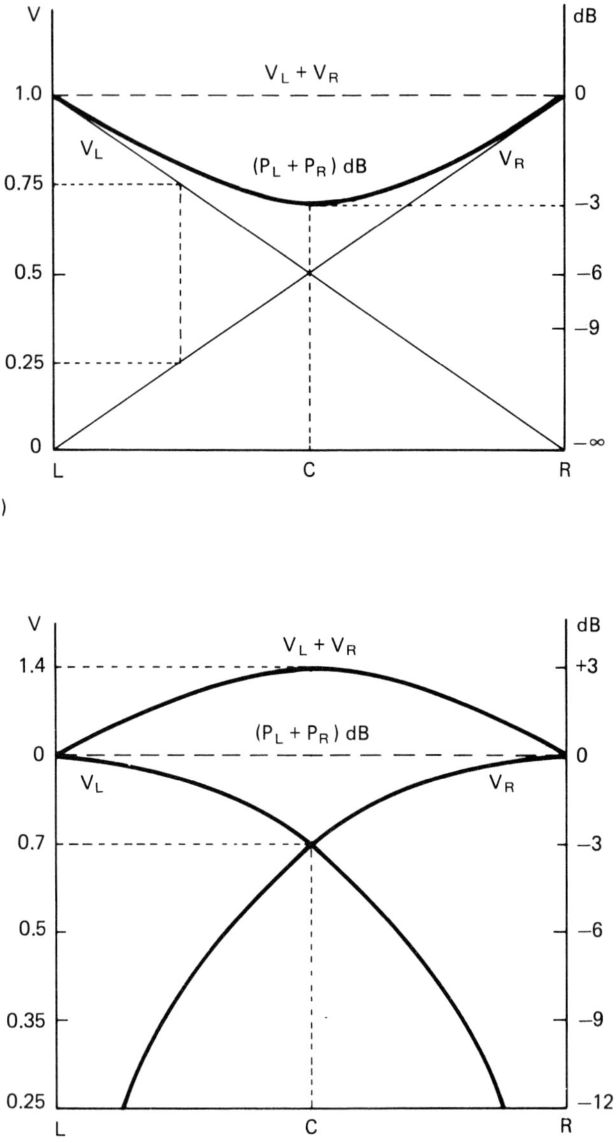


(a)

(b)

**Figure 1.5** If a signal is panned between left and right such that the sum of left and right signal voltages remains constant for mono, the sum of power for stereo will fall in the centre (a). If the power is held constant (b), the mono mix level will rise in the centre. At (c) a compromise curve (iii), lying between constant power (i) and constant voltage (ii) results in a small centre level rise in mono and fall in stereo.

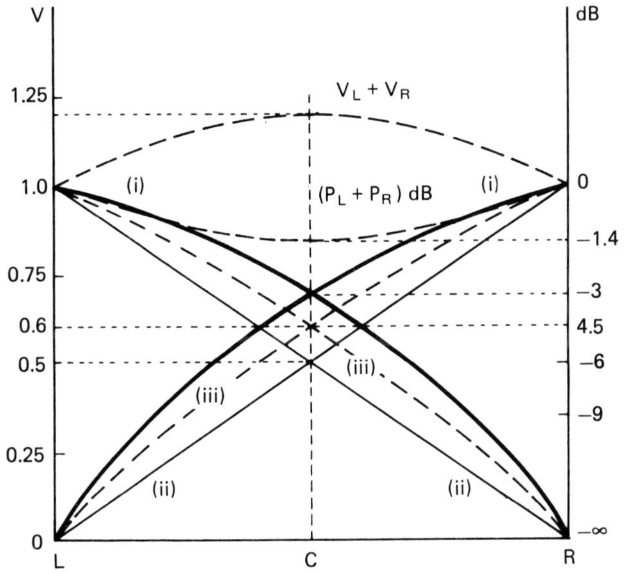


(c)

**Figure 1.5** *cont.*

example the SQN-4S location sound mixer can be used to process an MS microphone, but has the facility to switch the headphone monitoring to AB to permit proper assessment of the stereo image by the recordist. Mathematically the process of converting from one to the other is simple:

$$
\begin{aligned}
M &= (A + B)/2, \; S = (A - B)/2 \text{ so,} \\
A' &= M + S \\
&= (A + B)/2 + (A - B)/2 \\
&= A/2 + B/2 + A/2 - B/2 \\
&= A
\end{aligned}
$$

(Where A′ is the derived A signal)

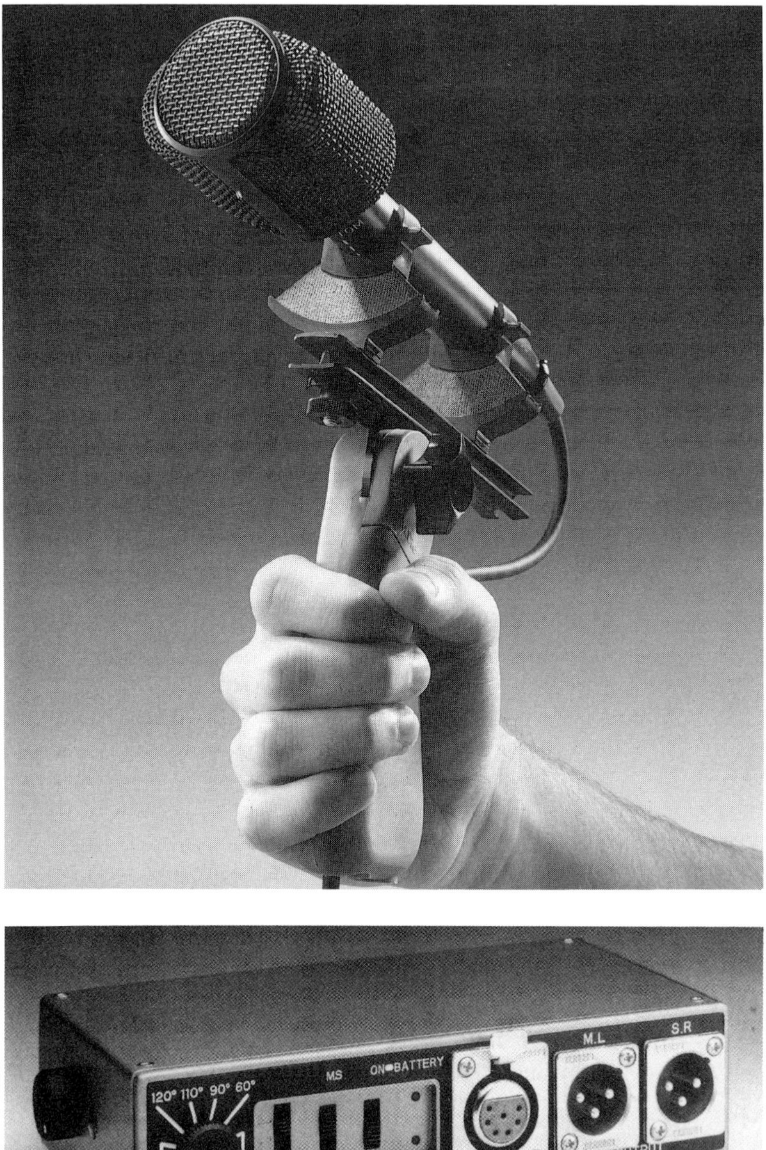


**Figure 1.6** A typical stereo microphone with its control box (Courtesy: Sanken).

Similarly

$$
\begin{aligned}
B' &= M - S \\
&= (A + B)/2 - (A - B)/2 \\
&= A/2 + B/2 - A/2 + B/2 \\
&= B
\end{aligned}
$$

(Where B′ is the derived B signal)

In practice it is simple to derive them electronically. Figure 1.7 shows two possible ways of converting from AB to MS. The same matrices will convert from MS to AB. The reversed winding on the B signal transformer, and the inverting amplifier in the B signal feed to the S summing amplifier, both provide the −B signal from which the S signal may be derived by summing with A.

## *Effects of phase inversion*

It is very important that the correct phase relationship between AB and MS signals is maintained at all times. Phase reversal, and phase shifts, have profound, though different, effects on both the mono and stereo images.

To take the AB system first: any phase reversal between the channels will cause varying degrees of cancellation in the mono sound. Signals placed in the centre will be affected most, and the result for the listener in mono will be that primary sound sources, which are usually placed in the centre, will appear to be considerably off-mic, probably to such an extent that the sound will be unusable. To provide for this eventuality, which can be caused by an incorrectly-wired microphone cable among other things, location mixers usually have the facility to reverse the phase on one channel (usually the right, though as absolute phase is not important it does not matter which of the two stereo channels has the phase correction facility). The wise sound recordist will, however, have a phase reversing lead to hand 'just in case'.

The effect of AB phase reversal for the stereo listener is more subtle. The stereo image appears wider than intended, even to the extent that the sound images can appear to extend beyond the baseline between the speakers, and under some circumstances can appear to come from behind or within the head of the

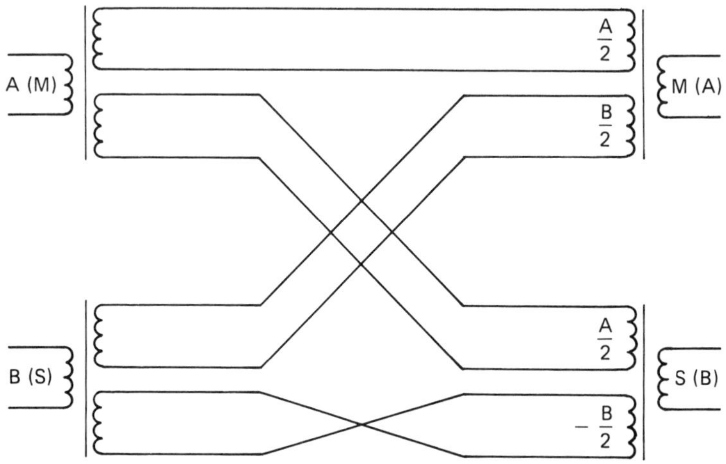


(a)

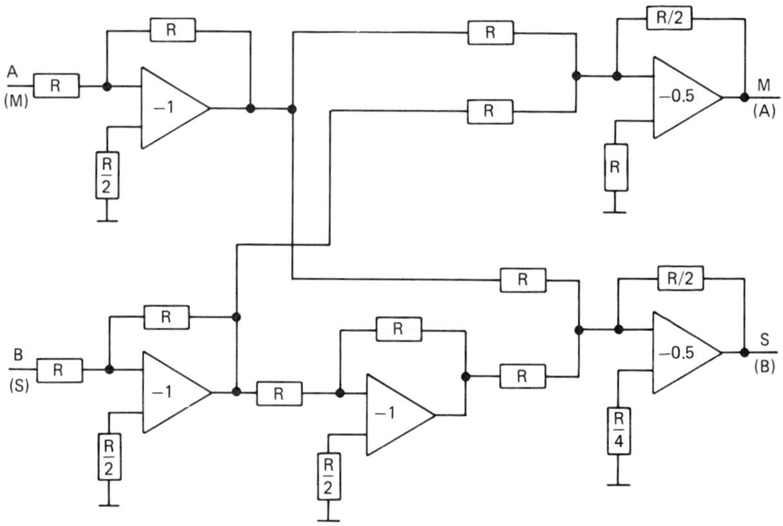


(b)

**Figure 1.7** (a) Transformers can be used to matrix between AB and MS. To minimize crosstalk and loading they should be hybrids. Operational amplifiers (b) can also provide the necessary matrixing, but if complex circuitry is to be avoided will require input and output transformers (not shown). Note that the same matrix will convert AB to MS and *vice versa*.

listener. Centre-placed images will be hard to locate; indeed they will usually be unlocatable.

Since a signal placed substantially to one side has the bulk of its energy in one channel only, the effects of phase reversal on an AB system are less noticeable for sound sources placed to one side.

Where MS systems are concerned, phase inversion between the M and S signals has no effect on the mono signal, since the S signal is not processed by a mono system. The effect of phase reversal on the stereo image is to reverse left and right when the signal is matrixed back to A and B. This effect can be put to great use in location recording, where reverse angles may have to be accommodated, or where the presenter may be holding the (end-fire) microphone left/right-reversed. And, of course, a pole operator frequently has to reposition the microphone from above to below and *vice-versa*, an operation which, of itself, will result in left/right image reversal. More than one prestige drama has been transmitted with the occasional scene left/right reversed. MS mixing desks and MS microphones usually have a phase-reversal facility provided for this purpose.

## *Interchannel differences*

Interchannel differences fall into three broad categories: level, frequency response and phase or timing (other than phase reversal). As with phase inversion, the effects of mismatched channels differ depending on whether we are considering an AB or MS system.

Taking the AB system first, level differences will cause the image to shift towards the channel with the higher level. Centre images will be placed over to one side to a degree dependent on the amount of level mismatch. The effect is most pronounced for centre-placed images, least for images panned hard to one side. For the mono listener there may be some perception of an incorrect balance, but this is likely to be unnoticeable.

Frequency response differences in the AB system will cause the stereo image to wander as the frequency content of the sound varies, and the image formed will be less well defined than if the channels are matched. Again, the effect for the mono listener is unlikely to be noticeable.

Errors in phase are a considerable problem for both mono and stereo reproduction, and it is important that they are not introduced into a system as they cannot be corrected at a later stage in production or transmission. The effects in an AB system for a stereo listener are most pronounced for centre-placed images, and result in a diffuse sound. For the mono listener the results are often disastrous, causing an effect known as comb-filtering (Figure 1.8). The perceived effect is impossible to describe in words. It resembles 'flanging', but once heard it is not forgotten.

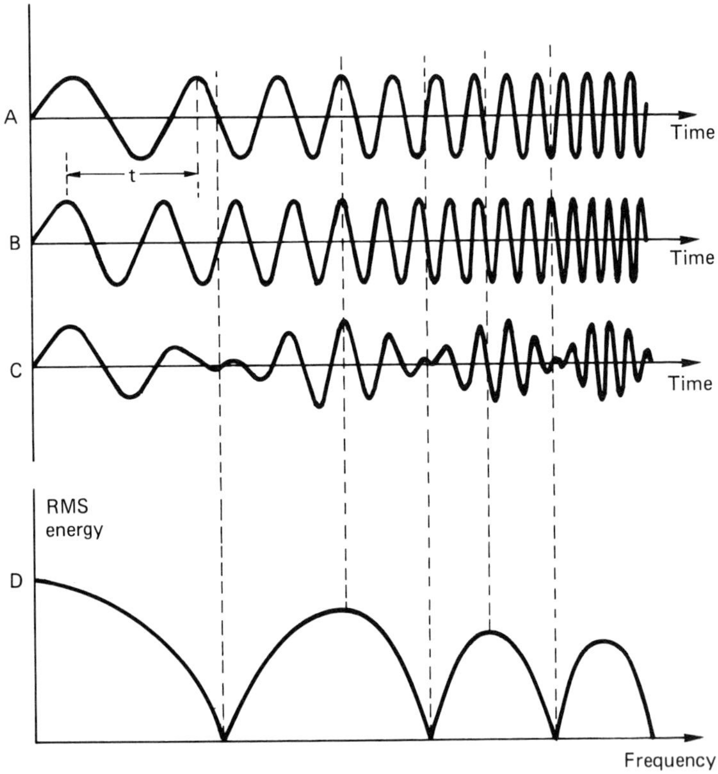


**Figure 1.8** Identical signals at (A) and, delayed, at (B) when combined will exhibit a complex phase-dependent relationship. In-phase components will reinforce each other, and those out of phase will cancel (C). The resultant energy spectrum with peaks and nulls (D) is called 'comb-filtering'.

Working with an MS system, variations in level or frequency response are unnoticeable in mono, where only the M signal is transmitted, but the effects on the stereo image cause variations in the width of the perceived image, with minimal effect on sources placed to the centre of the sound stage. However, differences in phase – particularly for sources placed to the side of the sound stage – cause very disturbing effects for the stereo listener. In an MS system, a source panned to one side appears equally on both M and S channels. When these are matrixed back to A and B form for monitoring and transmission, comb-filtering occurs. MS phase differences are not perceived by the mono listener as the S signal is not transmitted in mono.

### *Line-up tones*

There is no agreed standard between different organizations regarding the levels of stereo line-up tones, a fact not surprising since there is a multiplicity of 'standards' regarding mono line-up tones. However some principles can be considered. Firstly the tone levels fed to left and right channels must be identical in level, frequency and phase, so tone must come from a common generator and it cannot be recorded first on one channel, then on the other.

In the United Kingdom, the Independent Broadcasting Authority and the member-companies of the Independent Television Association require tone on the left-hand channel to be interrupted momentarily every three seconds. The reason for this is to identify track 1 of a tape as being the left signal of a stereo pair and not simply the mono mix: an important requirement when tapes are likely to pass through a number of different hands (not to say organizations) in the programme-making process. Not all location equipment provides this facility. It is therefore important that verbal identification is given, and given from the point of view of the camera and NOT the presenter, who may well be facing the other way!

As regards levels, there has been much debate about the relationship between the stereo signal levels and the derived mono signal. Probably the best advice is to record tones at 0 dB on both the stereo tracks and the derived mono track AND ENSURE THAT A WRITTEN LOG OF THE LEVELS IS KEPT WITH THE TAPE.

### *Dynamic gain control*

Very often some form of dynamic processing is employed on the longitudinal audio tracks of a camcorder in order to minimize tape noise (hiss, print-through). These processors have an effect on the recorded signal levels so it is important that, if noise reduction is in use, it should be switched IN when line-up tone is recorded. Again a note should be made on the log that noise reduction is being employed. Of equal importance is that the dynamics of analogue tape noise-reduction devices are matched both on the record and playback machines. If not, then any mis-tracking in the dynamic gain will cause image wander.

Another device commonly employed in the field is the compressor/limiter. Although most sound recordists regard this purely as a backstop in the event of sudden and unpredictable variations in level, it can nevertheless have an effect on the stereo image. Devices that are not ganged can cause the image to shift when, for instance, an off-centre signal causes only one of the pair to operate. Devices (usually 'off-board' in the context of portable single camera working) having differing thresholds, compression ratios and release times may also cause the image to wander even if they are ganged.

# 2

# Stereo microphone theory

## Monophonic compatibility

Any person involved in the recording, post-production and transmission of stereo sound for television must bear in mind that the signal will also be presented to many listeners in mono. It can be argued that in some respects the maintenance of a good monophonic signal in a stereo broadcast is more important than a good monochromatic picture in a colour broadcast, the rationale being that for many years to come the majority of viewers are likely to be listening in mono, and that many errors of interchannel balance, particularly as regards phase, are likely to have very disturbing effects for the monophonic listener. On the other hand, there is a considerable degree of evidence that demand for television sound obtained from a pair of spaced, high-quality loudspeakers, fed via a hi-fi system, is on the increase, even among viewers who would not previously have been thought of as hi-fi enthusiasts.

As far as the location recordist is concerned the rules for obtaining compatibility are few but important. Firstly, the way in which stereo microphones achieve their stereo properties can result in the introduction of unwanted phase reversals for sounds that come from too far round to the side (Figure 2.1). Moral: do not place the microphone in the middle of a field of sound, but instead try to place it in such a way that it looks at the field with the main sound sources being to the front.

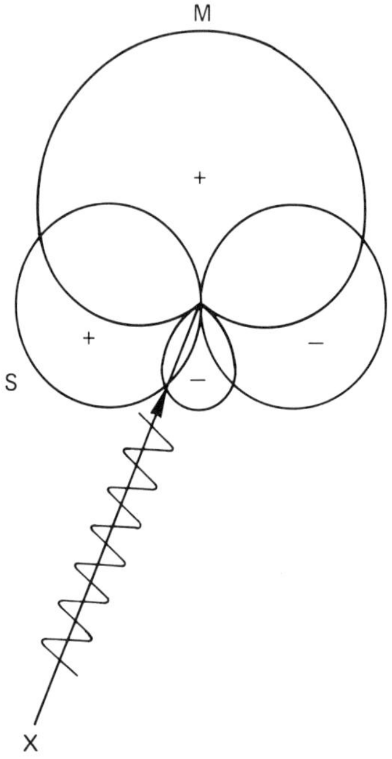


**Figure 2.1** A source too far round to the side of an MS pair may be subject to unexpected phase cancellation.

Some stereo microphones consist of two separate elements, which are physically clipped together by the recordist. If their relative spacing is not correct as specified by the manufacturer, then comb-filtering will result. When working AB this will be most noticeable for sounds located towards the centre of the stereo picture, that is to say the only sound obtained by the person with the monophonic receiver.

There will often be a requirement to provide a pair of stereo effects tracks, together with either mono or stereo commentary tracks. Separate microphones will be used for the two types of track configuration, and they are unlikely to be particularly near each other – certainly not approximating to the same point in space. As a result, sound common to both of them is likely to arrive at different times. If the sound reaching the microphones is comparable in level then considerable comb-filtering will result, with attendant degradation of both mono and stereo images. The answer to this problem is either to combine microphone pairs whose polar responses tend towards mutual exclusivity in the sound field (e.g. a hypercardioid with a figure-of-eight), or to space the microphones apart (e.g. to use a rifle with a stereo crossed pair): Figure 2.2 illustrates these possibilities.

The elegant solution of using non-coincident pairs and introducing electronic time delays is unlikely to be appropriate to the person working with portable single camera (and in the case of some newsgathering operations, portable single person). The separate effects and commentary tracks will often be recorded on physically separate tracks of the recorder. It is important that the combined mix is monitored in mono as well as in stereo to ensure compatibility.

Apart from mono compatibility, another reason for monitoring the combined mix is that Audio Channels 3 and 4 on camcorders are recorded as a part of the video signal, by the video heads, whereas Audio Channels 1 and 2 are recorded as longitudinal tracks, alongside, but separate from, the video tracks, using record/replay heads at some distance from the

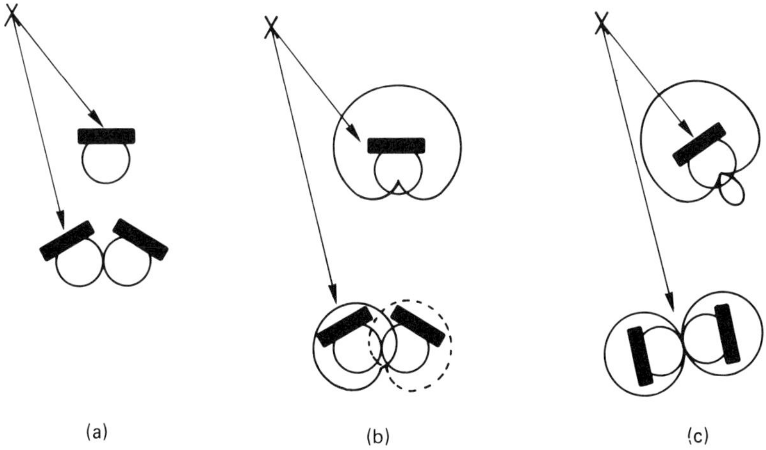


**Figure 2.2** If microphones are not coincident, sound from a single source will arrive at different times (a). If these spaced microphones are sensitive in the direction of the source (b), comb-filtering (and flanging if there is relative movement) may occur. When spaced microphones are positioned so the 'dead' axis of each is angled along the 'live' axis of the other (c), phase problems are minimized. However, the combined result is not true stereo since neither is picking up sound intended for the other.

video heads. When the tape is edited, the editing machine may very well not have exactly the same spacing between the longitudinal and video heads as the recorder. Although this difference will be of no consequence as regards lip-synchronization or with sounds that are substantially non-coherent, there will be phase errors introduced if substantially similar signals are mixed from the different tracks.

Another point to bear in mind is that the stereo effect will spread the various sound sources between the reproducing loudspeakers, and the ability of the human ear to distinguish between unrelated sound in space, the 'cocktail party effect', will allow selection of the wanted elements of the sound picture. The listener in mono will not have this facility and so positive discrimination in favour of the primary sources will be called for if the mono image is not to appear unduly confusing. Fortunately the physical laws that come into play when mono signals are derived from stereo pairs result in sound sources towards the sides of the stereo picture appearing lower in the derived mono mix. This could be taken as one more reason why prime sound sources should be placed towards the centre of the stereo image.

## Relationship between sound and vision

Most television sets have relatively small screens compared to the spacing of stereo speakers in hi-fi systems, and it is generally anticipated that, initially at least, most people listening to stereo sound on television will do so on their hi-fi speakers – if only to take advantage of the superior quality afforded by the digital transmission system employed. This begs the question of what the ideal relationship between picture size/position and sound stage width is. The problem does not arise in the cinema to anywhere near the same extent as the screen size is so much wider, a greater proporttion of the audience is likely to be in an ideal, or near ideal, listening position, and the Dolby System employed by the cinema for stereophonic sound is able to position centre signals firmly in the centre by the use of a complex matrixing arrangement, coupled with an additional loudspeaker placed centre screen. This is unlikely to be the case in the average television-viewing environment, be it in the home or the production control room, unless viewing a Dolby-encoded video via a properly installed Dolby decoder (domestic versions exist). Indeed it is quite probable that, in many cases, the screen will not even be central between the speakers.

Clearly some compromise will have to be accepted, and skills learned, to ensure as wide a range of compatibility as possible between the different domestic set-ups. Precisely what will ultimately be found acceptable is largely going to be a matter of personal preference on the part of the viewer; however some guidelines can be established.

It can be argued that the television screen, small though it is, shows us just a small part of an overall much larger picture, and it is certainly a fact that in real life we have a very limited angle of high definition vision. Good production techniques in both sound and vision, which start by establishing the overall location, perhaps by the use of an opening long shot accompanied by a wide perspective sound, will set the scene. Commentary can then be placed at centre, or near centre, and close to without distracting the viewer.

It can also be argued that stereo sound can be used to suggest off-screen action. If we accept this then we can say that sound and picture perspectives do not have to be inextricably linked. Most viewers of stereo television find that hearing sounds coming from places where there is no picture, although initially disturbing, is perfectly acceptable after a short period of acclimatization.

It is usually difficult to align sound sources with their visual couterparts so most organizations (including film makers) record dialogue and commentary in mono, and place it either firmly in the centre, or else pan it only slightly left or right of centre so that it still appears to come from within the confines of the screen for anyone viewing from a central position. This avoids the need for lengthy post-production time. The accompanying effects need not even be either 'true stereo' or placed accurately in the supposed visual panorama as long as obvious contradictions are avoided, such as footsteps going from left to right when their owner is moving in the opposite direction, or effects planted firmly to one side when their source is in close up on the screen.

One technique that is best avoided is to place relating sources of sound too far to the sides. For example, when covering a discussion between more than two people, resist the temptation to place the people seated on the outside towards the sides of the stereo image. It is a technique that works well in radio, but sounds totally unrealistic in television when the camera cuts to a close-up or mid-shot of the person concerned (Figure 2.3).

Sources of stereo sound that pan around, whether or not the camera pans, are generally found disconcerting so it is generally accepted that, unless there is a clear need for the sound stage as a whole to move, it is best to take it from a sound viewpoint that

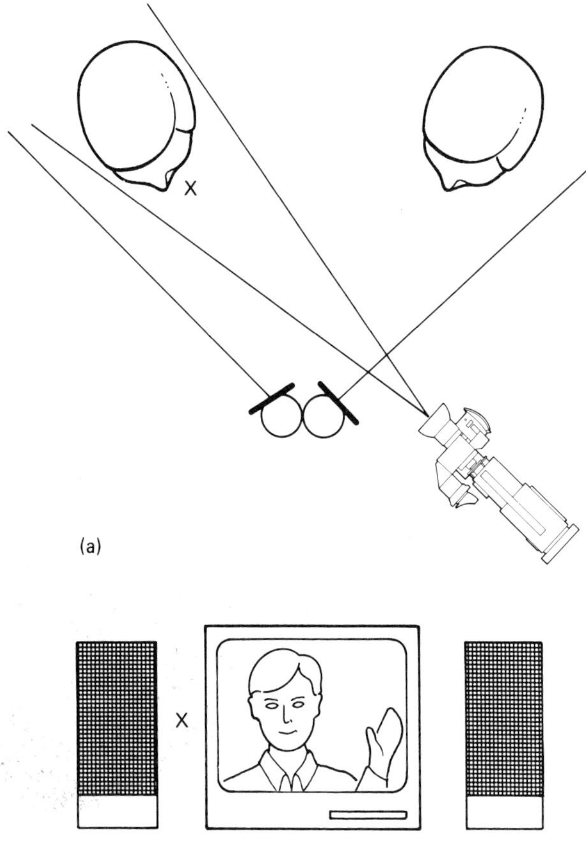


**Figure 2.3** The camera has the person of interest in frame. The microphone is angled to pick up both people (a). The result is a visual image firmly centred in the screen, but a sound image off to one side (b).

remains fixed. This is even the case for camera reverse angles, as long as they are not of too long a duration, and in this respect thought needs to be given to production technique – for example, tennis at Wimbledon is now shot predominantly from the same viewpoint, namely from one end of the court. If reverse angles have to be sustained in drama, it is perhaps better to fade stereo dialogue to mono then fade back up to a reversed stereo picture, covering the join with stereo wildtrack rather than cut, and certainly rather than pan, the sound.

What happens when the camera moves? Should we move the sound to match the picture? Most people find a sudden shift of the sound stage much more disconcerting than picture cuts between close-ups and wide-angles, and find 'audio pans' distinctly unpleasant. After all, if we are watching an orchestra at a concert, or a game of football, we are used to directing our visual attention between the wide angle and the particular item we are interested in, without expecting that the sound we hear will change accordingly. In recordings of classical music there is also the question of the spatial positioning of the performers to consider. A soloist such as a singer may be placed to one side of an orchestra in reality, but what happens when the camera cuts to a close up? Experience suggests that the rules for discussion programmes also apply here. That is to say, place the soloist firmly in the centre to ensure a sound and picture match for single shots.

Where sporting events or other activities in the public domain are concerned, the camera viewpoints are likely to change abruptly depending on the nature of the action. When considering the matching of sound to picture in these situations, remember that the brain has a remarkable facility to accept short-term contradictions between sound and picture perspectives, so it is better to establish a clear sound picture and stay with it, rather than attempt to match the sound to varying camera angles of view and perspectives.

## *Out of vision sound sources*

It has been argued in the previous section that sound sources from outside the picture can be perfectly acceptable, and indeed may enhance the stereo effect. This argument only applies to those sounds which are relevant. Extraneous and unwanted sounds are going to be much more intrusive firstly because of our ability to locate their spatial position, coupled with the fact that their position may move between takes/scenes – resulting in distinctly unnerving jumps in the stereo image to be accommodated during post-production, and secondly because the high quality digital sound played through a hi-fi system (likely to be used by the stereo viewer) is going to make them much more noticeable anyway.

There is also the matter of sounds from unseen sources that are unrelated to the event. Perhaps there is a bus, unseen but heard, off to one side. In mono working, the fact that all of the sound comes out of the same hole means that unwanted sounds are, to an extent, masked by those that we are interested in. This is not going to be the case in stereo.

Even wind and handling noise are going to be more of a problem in stereo than in mono simply because stereo microphones contain directional (figure-of-eight) elements to obtain the S component of the stereo signal and these polar responses tend to give microphones that are prone to interference from both these effects.

There are good reasons for taking more care over unwanted sounds in stereo than in mono.

## The stereo microphone

The stereo microphone usually consists of two elements in a single housing. They are designed to be as close together as possible to minimize phase difference effects (Figure 2.4). Each element consists of a directional microphone whose relative sensitivity may be altered, usually by means of a multi-position switch. In general a stereo microphone will be designed to provide either left and right signals, AB working, or sum and difference signals, MS working. There is an increasing number of microphones available that may be switched between AB and MS.

Microphones may have their two elements mounted either in separate capsules, one above the other to permit the acceptance angle to be varied by rotating the capsules relative to each other, or the two elements may be fixed relative to each other, with an

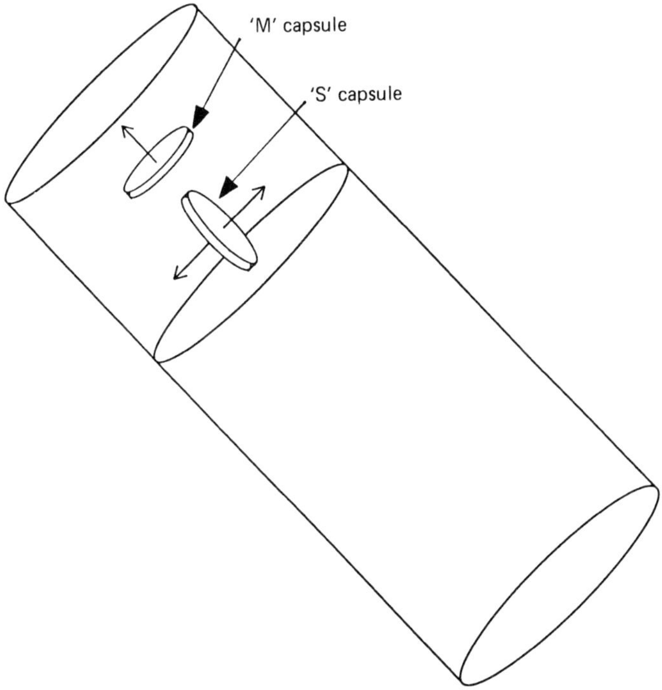


**Figure 2.4** The capsules of this MS microphone cannot occupy the same point in space. Manufacturers aim to minimize the phase effects this can cause.

electronic width control. One thing is certain, if a microphone is to be of much use to the sound recordist working in the field then some form of adjustment over the acceptance angle is essential. Moving the microphone nearer or further from the action to get the correct audio image width is at least as unacceptable as when working in mono, and many would say more so.

The capsules of 'side-fire' microphones (Figure 2.5) may be switchable between omni-directional, cardioid and figure-of-eight responses. When the elements are switched to omni then no directional information is obtained. With both elements at figure-of-eight, the microphone will supply a stereo picture, but sources at 90° to the side will appear as though there was a phase reversal between left and right. Sources to the rear will appear in phase but left/right reversed.

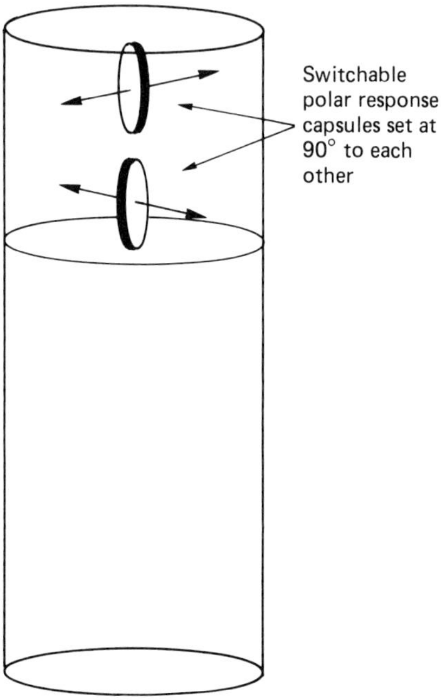


**Figure 2.5** The capsules of side-fire microphones are angled with respect to each other. They can frequently be rotated relative to each other, though the alignment of the individual capsules is not always clear from the outside, even when the manufacturer has provided markings. Their polar responses are generally switchable from omni, through cardioid and hypercardioid, to figure-of-eight.

Where 'end-fire' microphones are concerned the elements are either likely to consist of a pair of closely spaced cardioids (AB) (Figure 2.6), or else one of the elements will be a sideways-facing figure-of-eight, usually with its gain variable on a switch, coupled with a forward-facing directional microphone (MS), usually with a directional response lying somewhere between a cardioid and a hypercardioid. This has been found to give the best compromise between acceptance angle for monophonic compatibility, while not being too sensitive to sounds coming from the rear, and also keeping the out-of-phase region to a minimum. The polar response of the forward-facing capsule is not usually switchable, though one manufacturer does make such a microphone, primarily for use on studio sound booms.

Making the gain variable on the figure-of-eight element gives control over the stereo acceptance angle. As the S channel gain is increased, the fully left and fully right points move inwards. This has the effect of narrowing the in-phase acceptance angle BUT WIDENING THE STEREO PICTURE (Figure 5.1, page 96). Conversely, decreasing the S channel gain widens the in-phase acceptance angle and narrows the stereo picture.

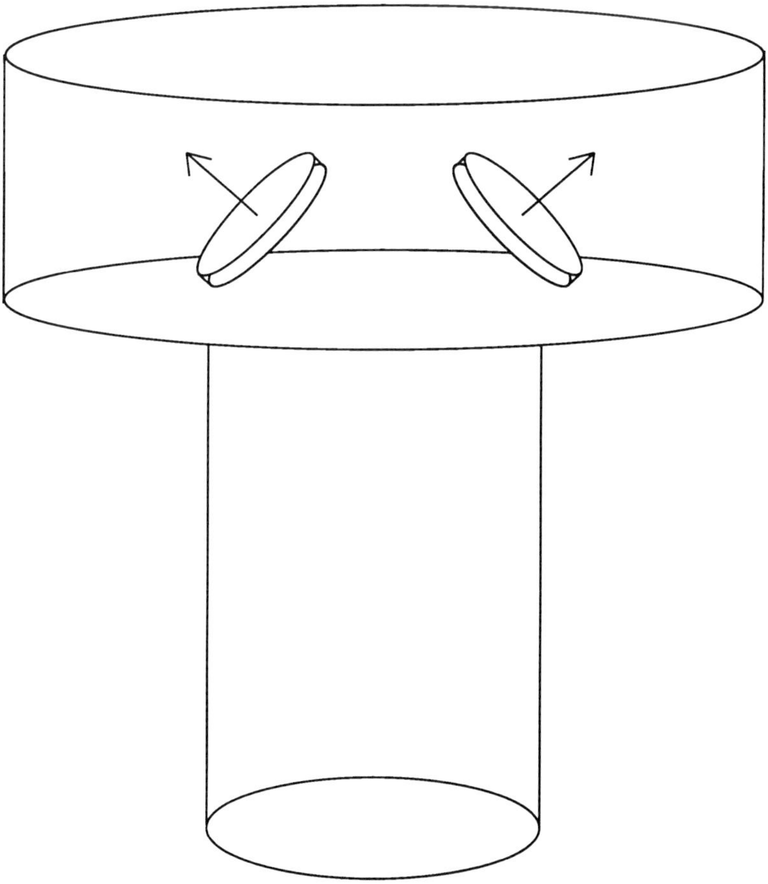

**Figure 2.6** Some manufacturers of end-fire microphones use left/right facing capsules, and provide a control box to vary both polar responses and the MS ratio.

## Mono plus ambience techniques

Some recordists have reported successes using what might be termed 'pseudo stereo', or mono plus ambience techniques. In these cases the basis has been to work with sum and difference microphones, but not to have them coincident. The justification for this way of working is that in some situations it is impossible to place a rather bulky stereo microphone in position. Instead, the M component is replaced by a discrete mono microphone, such as a personal or small stand-mounted microphone placed in the optimum position, and to place somewhere near, but out of shot, a figure-of-eight mic to provide 'stereo ambience'.

As long as there are no clearly defined sound sources to either side of the stereo picture, and the mono mic is on one of the dead sides of the figure-of-eight, this technique can provide surprisingly good results (Figure 2.7).

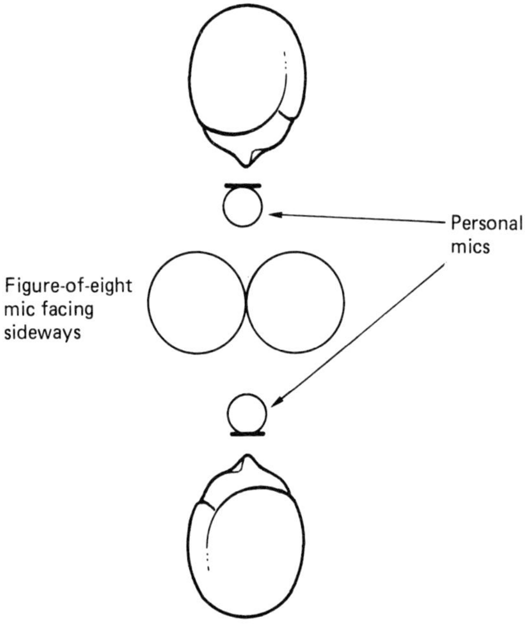


**Figure 2.7** If a pair of personal microphones are panned to the centre of a stereo mix, and a figure-of-eight fed both left and (phase inverted) right, the result will be well defined centre images with a good 'stereo ambience'. There should be at least 1 m spacing between each of the personal microphones if flanging effects are to be avoided.

Success in the recording of choral music has also been achieved using a combination of personal microphones on the soloists and crossed stereo pairs on the congregation.

## Multiple microphone techniques

To obtain a stereo image from a number of mono (spot) microphones, their outputs are panned between left and right electronically, the variations in relative left/right levels being used to position the source between the speakers.

Many sound recordists already use multi-microphone balances in mono, and there is some justification for saying that multi-mic balances in stereo are less difficult, as the directional information gives the separation between sources that would otherwise have to be achieved by altering the balance between their relative levels, timbres and perspectives for mono.

There are, however, a couple of points to bear in mind, chief among these being that sources intended for one microphone should not be picked up to a substantial degree by any other (an effect called spill), otherwise phase errors and/or shifts in apparent image position will result.

A mix of panned mono microphones alone will not give an overall ambience, since to avoid spill microphones usually have to be placed close to their intended sound sources. In recording studios this shortfall is usually made good by adding artificial ambience from an electronic device. This is rarely satisfactory, or indeed possible on location shoots, so a stereo microphone, placed some distance back from the sources, is mixed in. It is important when mixing the output of this microphone in with the spot mics that the positions of its sound images in the stereo picture correspond to their positions on their appropriate panned mono mics.

# 3

# AB microphone techniques

## Introduction

As in mono, stereo recordists will spend a great deal of their time worrying about matching perspectives and shot changes. These are sometimes in conflict, so it is worth restating a basic ground rule in the perception of sound, namely: *we are more disturbed by changes in quality than absolute quality*. Of course the location sound must be good, but the definition of 'good' depends on many factors, the most important of which may well have more to do with the shot or the edit than with the sound quality. In many instances we will need to forgo stereo realism to make an edit acceptable.

Most of the sound on television is dialogue, and the majority of this will still need to come from the centre of the image. Therefore the question must be asked, 'what is the point of recording dialogue in stereo?' Most of you are probably aware that the film industry gave up that idea almost as soon as it started making stereo films.

The answer is: because it gives you a stereo ambience that is obviously correct for the environment and, let's face it, it's cheaper than dubbing it in later. Although at first glance this ambience might seem minor, it can sound extremely disconcerting to hear a mono recording in the midst of a stereo programme.

Stereo can be recorded in a number of different ways. You have to decide which is the most appropriate for your needs. Each technique has its own strengths and weaknesses. At its

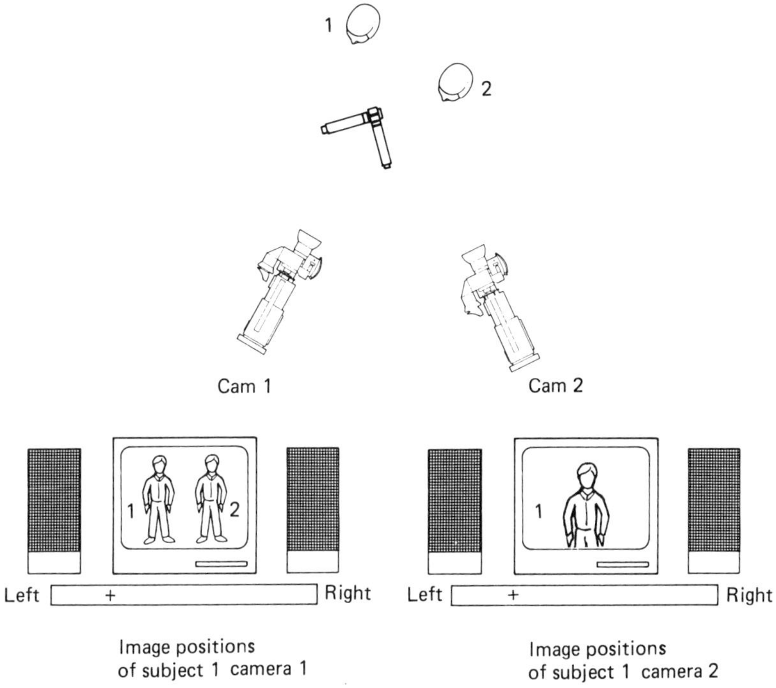


**Figure 3.1** A typical stereo problem. Sound and picture images may not match when the camera viewpoint changes.

most basic you have a choice between pair stereo or multi-microphone panned stereo. In parallel with this, you need to consider whether you would benefit from MS working.

However, we will begin with AB working. After all we ALWAYS listen to stereo as AB, so why not record it the same way? We can use most of our old trusted microphones and don't need a matrix to monitor the results! You have control of the level, balance and the width.

## Coincident pairs

The most simple and pure form of stereo is the coincident AB pair. It is the foundation of stereo theory. Coincident capsules also give you excellent mono compatibility, as they receive no timing differences regardless of the direction of the sound source.

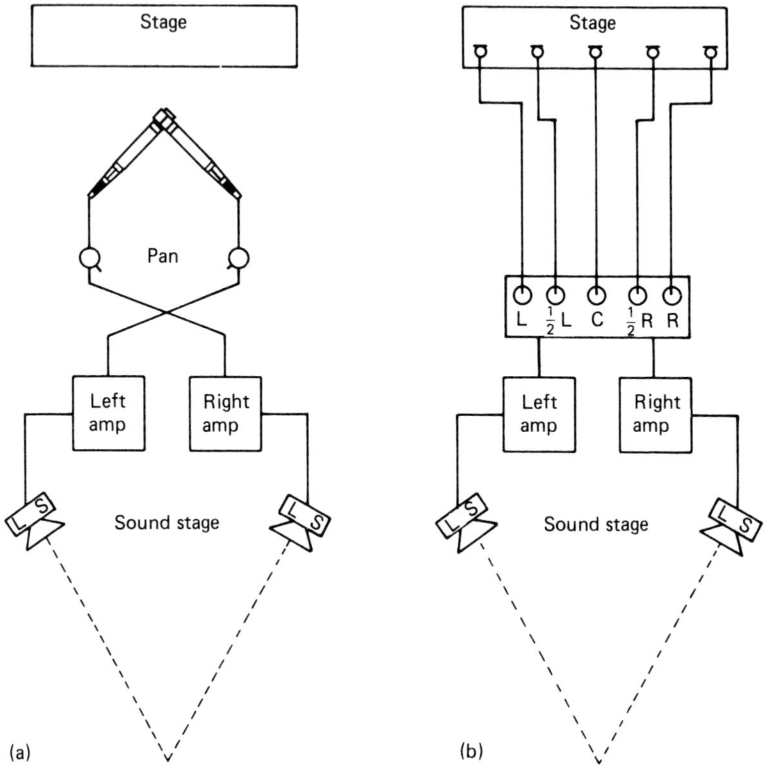


**Figure 3.2** Crossed pair stereo (a); Multi-mic stereo (b).

The two microphones you choose for your pair must be a very good match in all the usual parameters, including frequency response, polar pattern and sensitivity. Always use good quality microphones and give them a thorough listening test. Two microphones of the same model but of different ages may not be compatible enough to be used as a stereo pair.

Your first problem is to decide which polar patterns to use, as this has a profound effect on the resulting stereo.

### *Crossed figure-of-eights*

Crossed figure-of-eights at 90° give superb and accurate imaging across the full 90°. However, figure-of-eights do have well defined out-of-phase regions and any reflected sound that is picked up here will damage the stereo. Also, because there are equal and opposite rear lobes, the situation can arise when a sound that is, say, on the rear left of the microphone will in fact be picked up on the right capsule. This is not a problem if the rear pick-up is only ambience, for then the results can be very pleasing, especially on music recordings, but it can be most unpleasant if it is a primary source of sound.

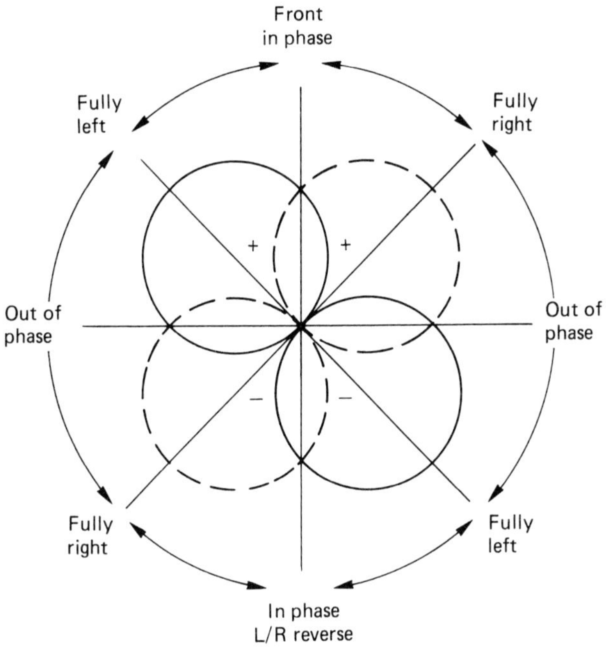


**Figure 3.3** Figure-of-eight microphones at 90°.

Another feature of crossed figure-of-eights is the 90° acceptance angle. Any attempt to widen the angle will result in a dead area right in the middle of the image. This is known, not surprisingly, as the hole-in-the-middle effect. Therefore to cover a wide sound stage, it is necessary to move a crossed figure-of-eight pair away from the action, instead of altering the acceptance angle. This may well result in the recording sounding too distant. Later we will see that changing to other polar patterns can give you increased acceptance angles.

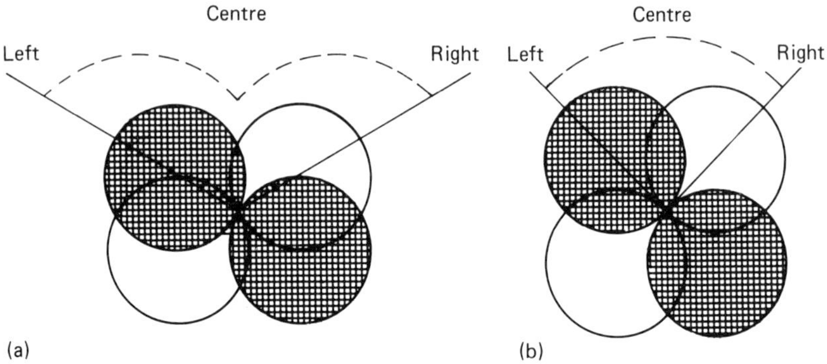


**Figure 3.4** Figure-of-eights at 120° (a) produce a hole in the centre of the image, and figure-of-eights at 90° (b) produce a linear sound stage.

When choosing the polar patterns it is useful to keep in mind the mono equivalents of the stereo pairs. For instance, crossed figure-of-eights give a very realistic stereo image but their derived mono is a forward-facing figure-of-eight. There are only a few special situations, other than music, where you would have covered a mono recording in this way. Therefore, despite their accurate stereo, you should only use crossed figure-of-eights after careful consideration of the resulting mono.

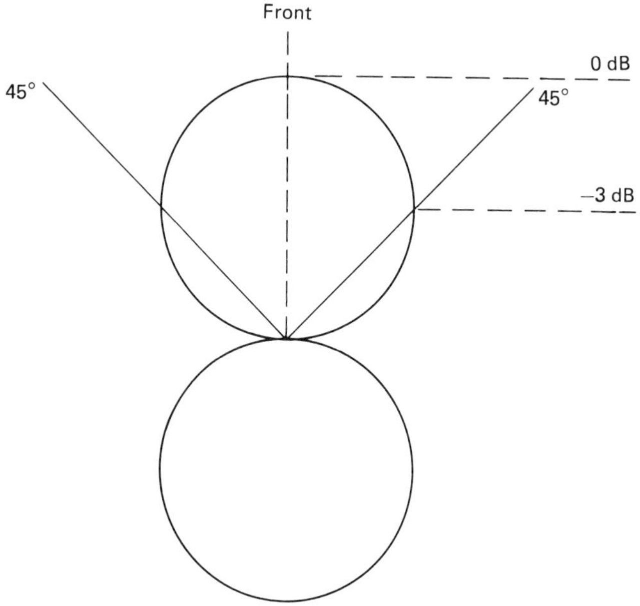


**Figure 3.5** Mono resultant for crossed figure-of-eights (90°).

*Crossed cardioids*

Crossed cardioids give a firm stereo image and the resultant mono is an acceptable if somewhat 'fat' cardioid. This configuration works best quite close to the subject, where you have some control over the background noise.

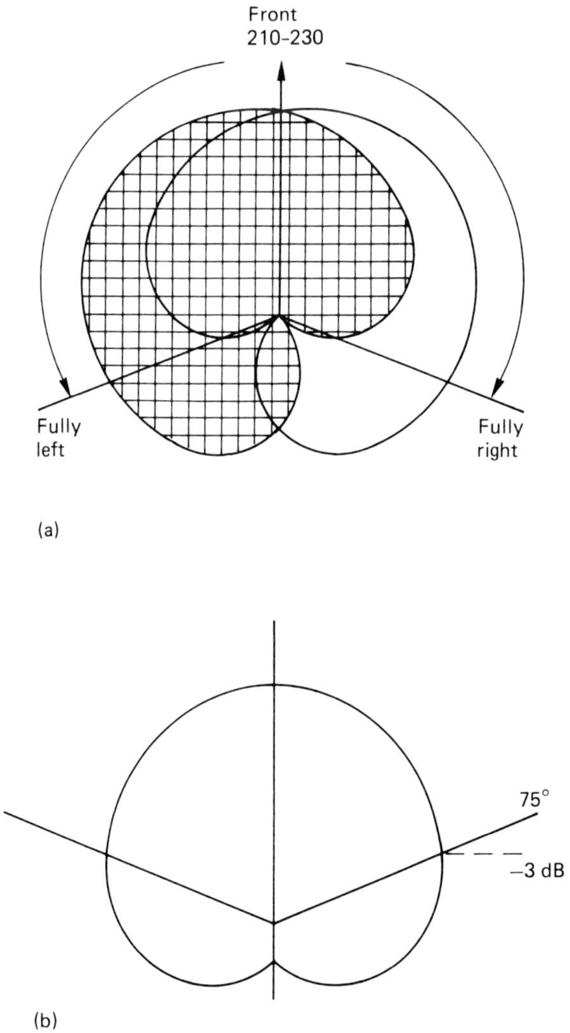


**Figure 3.6** Cardioids crossed at 90° (a); Mono resultant (b).

*Crossed hypercardioids*

If the shot pushes you more than a few feet from the action, most recordists favour moving to crossed hypercardioids. However, you must be aware that the angle setting between the microphones is more critical than with the cardioids. If the angle setting is too wide, you will deform the sound stage with a pronounced hole-in-the-middle. Hypercardioids do tighten up the derived mono but the rear pick-up seems to give a greater feeling of room acoustic.

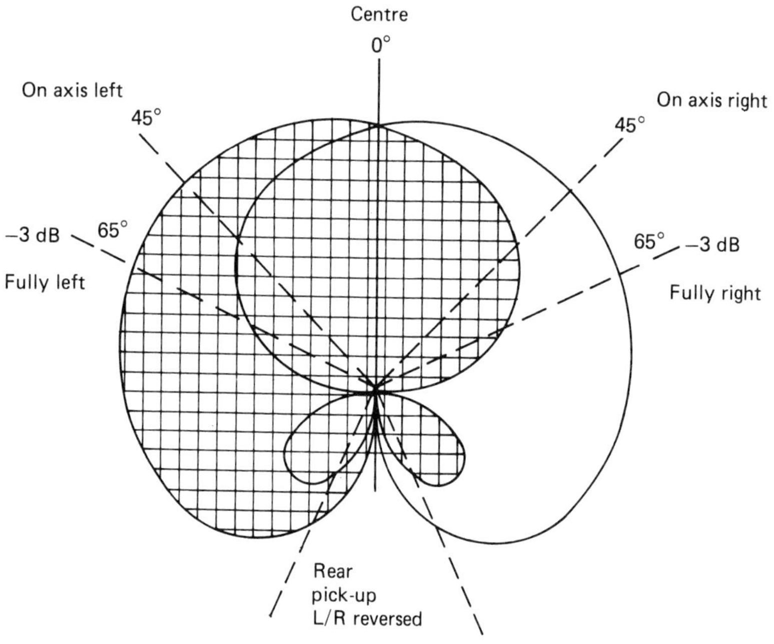


**Figure 3.7** Hypercardioids crossed at 90°. Centre is not on axis of the microphones. Edge images are up to −3dB down on central images.

It is important to appreciate that the setting of the capsules will have a major effect on both the stereo and the mono signals. Coincident really *means* coincident; a separation of only a few millimetres will introduce timing errors at high frequencies. These will cause ill-defined stereo images for central sources and high-frequency cancellation in mono.

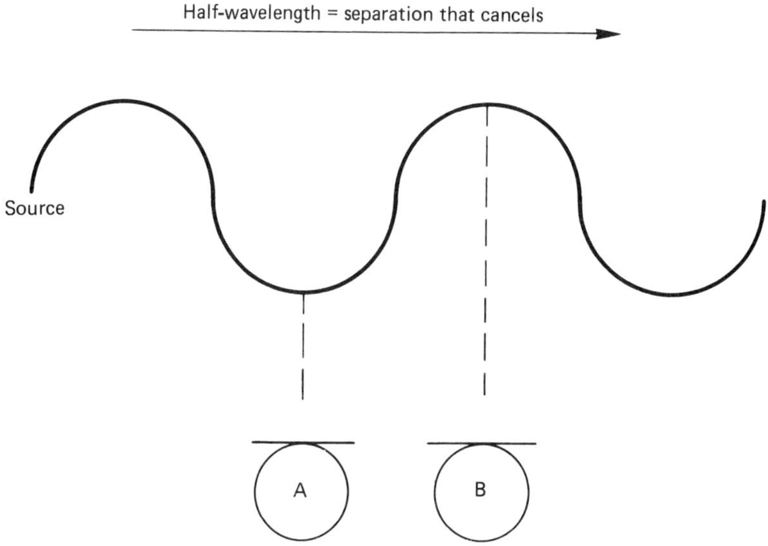


**Figure 3.8** At a capsule separation of only 2 cm, sources off to one side will cancel at 9.5 kHz, with phasing effects at lower frequencies.

### *Crossed shotguns*

It is worth reminding ourselves at this stage that for many, if not most, situations in mono we would probably have used a short shotgun microphone of the Sennheiser 415/6 type. Yet none of our stereo configurations gives us this level of directional accuracy. Crossed shotguns are an option, and provided the included angle is small, about 40°, and you remember to cross the tubes at the capsule, they will give an image. However, because of their method of operation, interference tube, the resulting stereo is notoriously unpredictable, as is the mono compatibility. This is doubly true with full shotguns of the 815/6 variety.

### *Back-to-back cardioids*

There is another coincident pair configuration, back-to-back cardioids. They are seldom used but do have some interesting characteristics. There are no out-of-phase areas and the

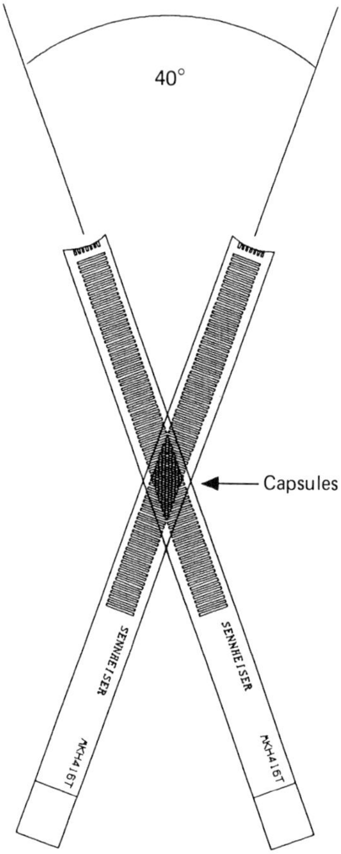


**Figure 3.9** Crossed shotguns.

acceptance angle is theoretically 360°. Unlike figure-of-eights, the rear pick-up is a mirror image, rather than a reverse, so walking around the pair does not flip the sound stage. The major disadvantage is that the mono resultant is omni-directional and this limits its use in television. However, the 'full view' stereo might be worth considering in some situations.

### *Choosing the acceptance angle*

The width of the stereo image between the loudspeakers is called the sound stage. You have the ability to compress or expand the

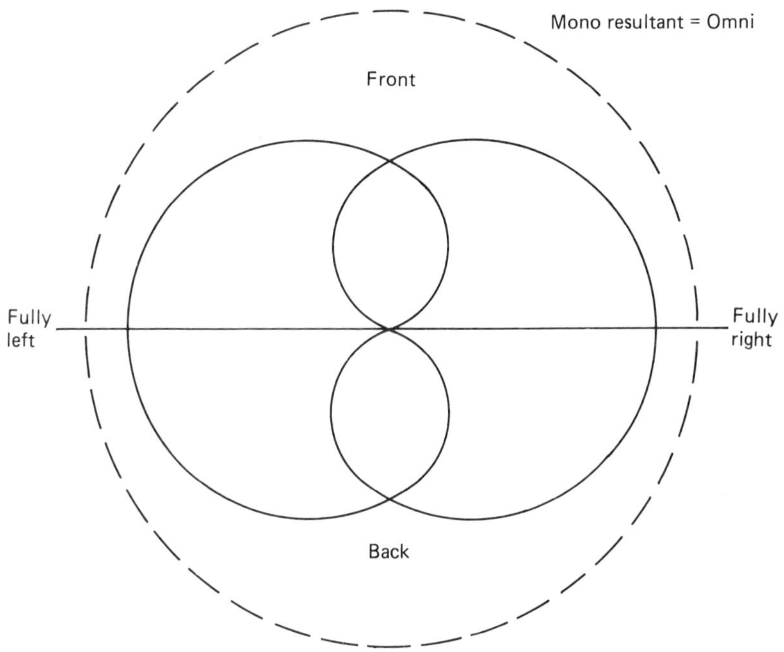


**Figure 3.10** Back-to-back cardioids. No out-of-phase area. No phase reverse.

image to fit the pictures. This is achieved by adjusting the angle between the microphones. You should avoid over-narrow or over-wide images unless you have very good reasons for them. There can sometimes be confusion about the relationship between the stereo width and the acceptance angle. This is very important and needs to be understood.

As you increase the acceptance angle, you in fact narrow the stereo sound stage, as it is now necessary to go further from the centre to become fully-left or fully-right. Fully-left or right is the point where the opposite capsule output is at a minimum and not the maximum on axis signal. This means that when positioning the capsules, you should be aware that their physical angle may be less than the actual sound stage you are recording. For example, hypercardioids crossed at 90° have an acceptance angle of around 140°.

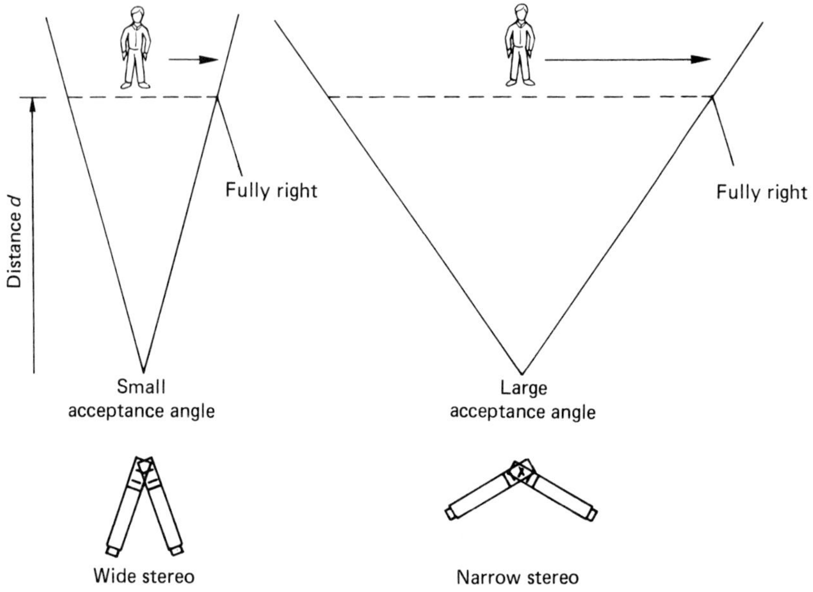


**Figure 3.11** Large acceptance angle = wide sound stage = narrow stereo.

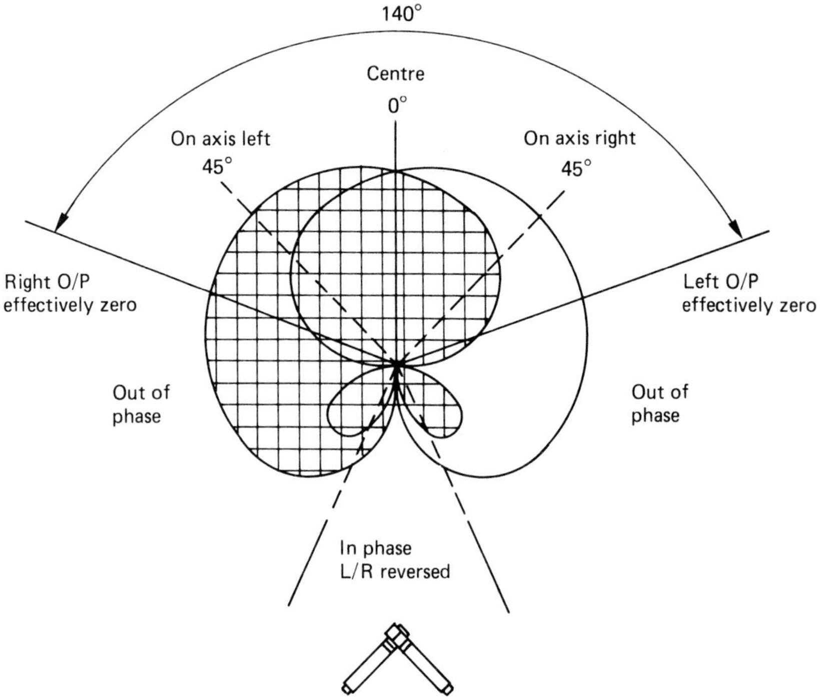


**Figure 3.12** Hypercardioids at 90° give acceptance angle of approximately 140°.

### *Acceptance angle and distance*

Another area of concern is the relationship between the acceptance angle of the microphones and the distance to the sound source. As you can see in Figure 3.13, as you move towards the subject the amount of movement needed to swing the image from fully-left to fully-right changes dramatically.

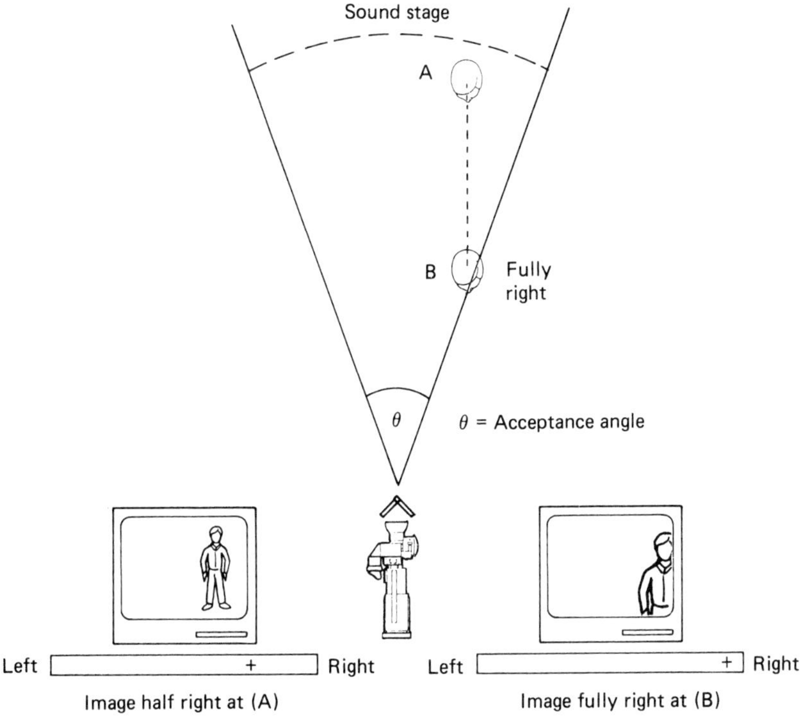


**Figure 3.13** Effect of distance on sound stage.

Also a source moving from point A to point B will move from half-right to fully-right, which would totally contradict the camera's view of the scene.

In a perfect world, as the distance from microphone to source decreases, you need to physically widen the acceptance angle, which would be almost impossible to achieve in practice. But if converted to MS stereo on location or in post-production, the excessive width can be reduced. Although collapsing the whole thing to mono is always an option, it is an admission of failure. Incorrect microphone placement cannot be fixed using MS techniques; in fact it will probably make things worse. Off-mic is always off-mic.

It is worth noting that physically altering the acceptance angle will change the mono pick-up of the pair, whilst MS mixing leaves the mono untouched.

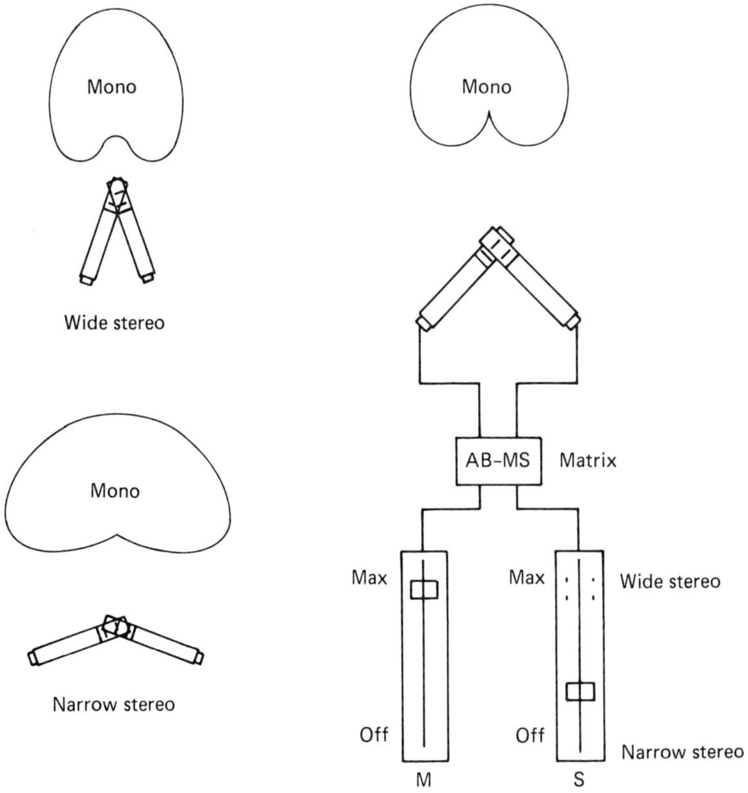


**Figure 3.14** Mono pickup alters (left); Mono pickup remains constant (right).

## *Calrec Soundfield*

The importance of coincident placing of the microphones cannot be overstressed. However, there is physically no way that two elements can be at precisely the same point in space. So true coincidence is unobtainable. In an effort to achieve the nearest thing to it, Calrec have designed and built the Soundfield microphone system. Using an array of four special capsules mounted in a tetrahedron together with some very clever electronics, they have produced an outstanding stereo microphone. The actual theoretical basis is quite complicated but can be best understood if you approach the microphone as if it were three figure-of-eight capsules mounted at the same point in space. Called X, Y and Z, they can be recorded directly to tape in the so-called B format mode, or pre-mixed to stereo. Adjusting the relative gains in the B format mode, either live or in post-production, can make the microphone appear to look in any direction horizontally and up to 45° vertically, with an acceptance angle of anything between 0° and 180°

Using the same tetrahedral capsule array Calrec have produced a more compact microphone for location work, the

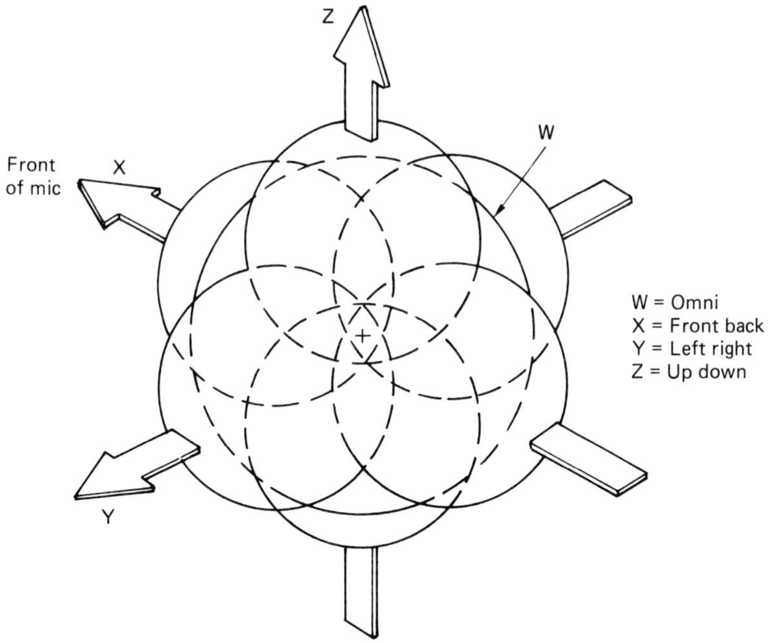


**Figure 3.15** The soundified microphone's tetrahedral capsules allow the soundstage to be steered either live or off-tape if recorded in B-format.

AMS ST250. The control box has been simplified and configured for the recordist. This is a truly unique microphone; not only does it offer the normal acceptance angle control but you can make it act like any type of crossed pair at any angle. The microphone can appear to be side-fire or end-fire depending on your requirements. It can output in true MS or AB. If you would rather leave the actual settings until later you can output in B format and electronically position the microphone in post-production!

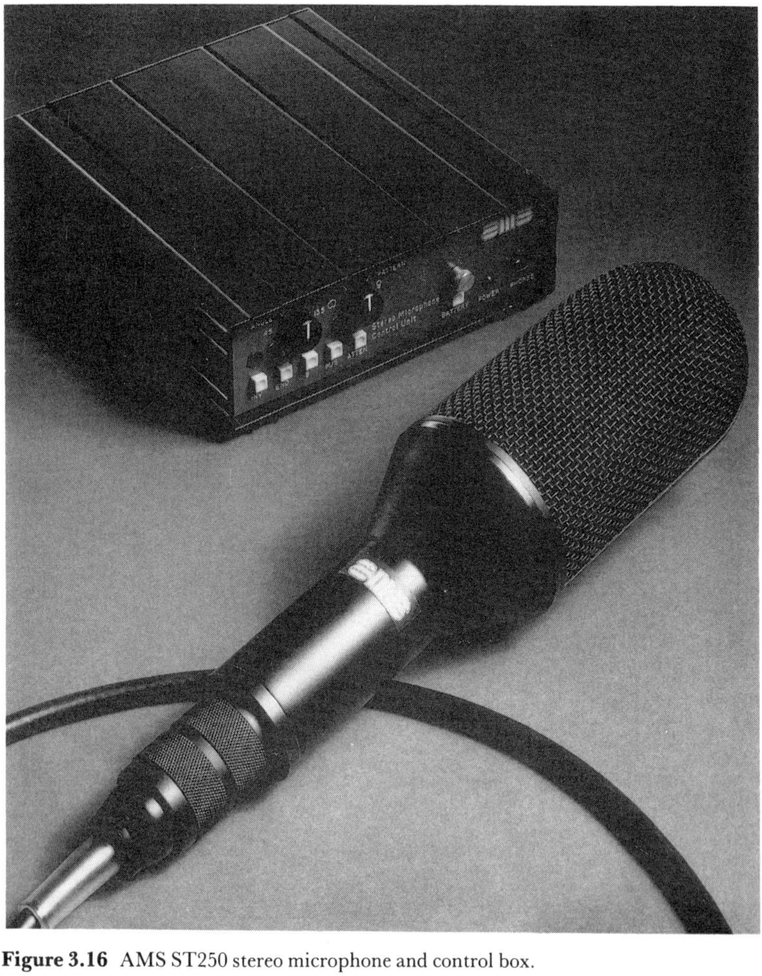

**Figure 3.16** AMS ST250 stereo microphone and control box.

## *Near-coincident pairs*

One of the major features of coincident miking is the absence of timing differences between the capsules. However if you do reproduce some timing differences by separating the microphones, you can get an increase in ambience.

In mono, the central sources are reproduced as before. The frequency-dependent phasing effects produced by the capsule spacing progressively increase as you travel across the sound stage. However these need not be a major problem. Provided you carefully check your mono compatibility, you should be able to achieve some very acceptable stereo.

Certain polar patterns are more suited to this technique than others. The most popular is probably crossed-cardioid as per the ORTF technique, i.e. 170 mm apart at 110°.

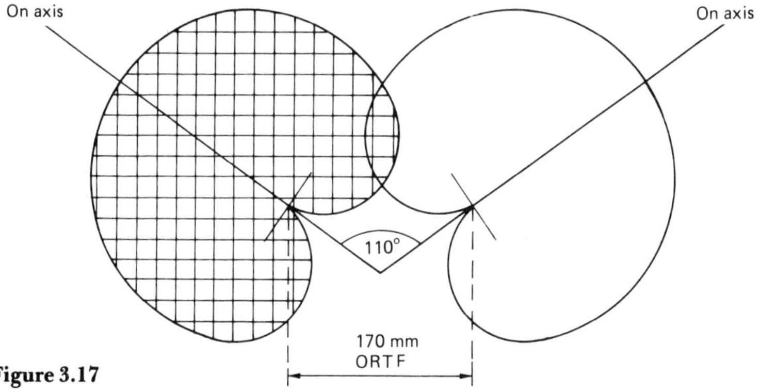


**Figure 3.17**

The stereo width of near-coincident pairs can be a bonus in 'dead' acoustics. But be aware that if you use a near-coincident pair close in, even small head movements will cause the image to jump across the sound stage.

## *Positioning fixed AB pairs*

The positioning of fixed AB pairs is critical but not difficult. The best method of choosing a site for the microphones is very basic. Walk around and use your ears! Choose a position that gives you a good naturally balanced sound field. You want to be far enough away to get a good perspective without too much room coloration, but not too close that you need an unusually large

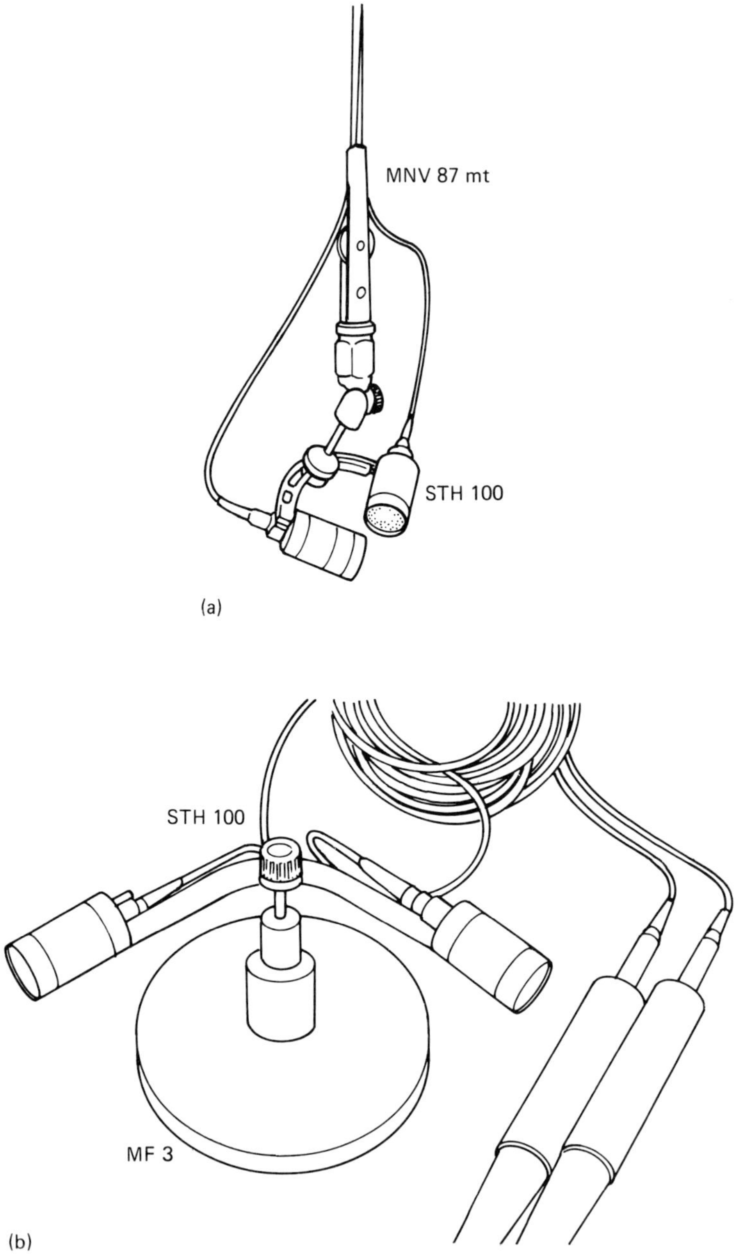


**Figure 3.18** Near coincident miking using active capsule microphones.

angle between the microphones to cover the available sources. It is worth keeping the pair as far away from reflecting surfaces as possible, so slung microphones cover a multitude of sins. But never overlook the fact that the sound perspective should match the camera shot. A good guide for placing static stereo microphones is to place them exactly where you would have put a mono microphone under the same circumstances.

Listening tests should convince you of the need for almost all stereo sound stages to be symmetrical. Irrespective of the visual cues given by the pictures, it quickly becomes irritating if there is an off-centre bias to the sound. The main pair should be placed to give you this overall stereo picture and you may add to, or enhance, it with other microphones. However, do not be tempted to try to fix an asymmetric sound field, caused by a poorly positioned pair, by using spot microphones. This is almost always unsatisfactory as it fails to consider the different perspectives of, or timing errors between, the microphones.

### *New microphones for AB stereo*

Many of the objections to the use of AB pairs revolve around their cumbersome appearance. However, in recent years, manufacturers have begun to offer a number of new breeds of microphone that are well suited to the needs of television stereo. Perhaps the most significant are the latest of the detachable-capsule capacitor microphones. The idea of detachable capsules has been around for some time but these offer new, very high quality, active capsules, designed to run on special low-loss cables. This makes them extremely neat and easy to position. Manufacturers also market a selection of stereo accessories for coincident and near-coincident techniques.

There is also a new miniature shotgun-type microphone, the AKG 747. This discrete microphone offers very good directional capabilities with a firm stereo image and yet is small enough to use in-shot or concealed close to the action. As with its larger kin, the acceptance angle should not exceed 40°, and the microphones need to be crossed at their capsules.

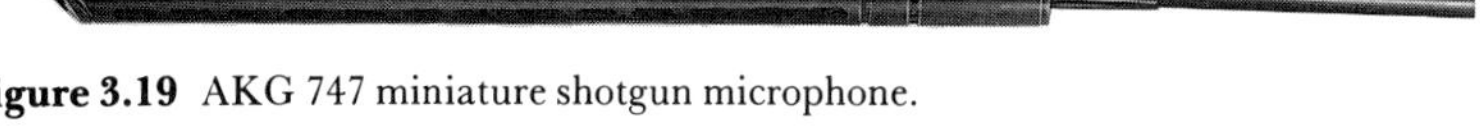


**Figure 3.19** AKG 747 miniature shotgun microphone.

## *Side-fire stereo microphones*

One-piece side-fire stereo microphones have been around for a long time and have a loyal following in radio and classical music recording. One capsule is mounted above the other and can be rotated with respect to it. This means that the two capsules are coincident to the sound stage. As fixed microphones, they are very reliable and easy to set up, as they come with multi-way cables, which aid rigging. However they can be fairly difficult to aim.

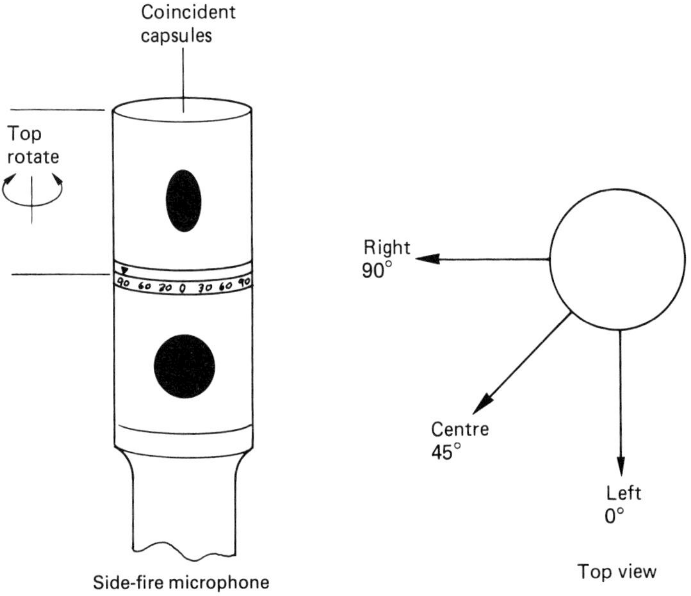


**Figure 3.20** Front of axis is on-axis capsule.

Most offer variable polar patterns which can be remotely controlled. This allows their use as either AB or MS pairs, although unless you have the latest versions, the angle markings on the bodies become totally meaningless when working MS! In an effort to simplify what should, after all, be a simple task, certain manufacturers now place LEDs on the forward face of the capsules, to allow quick and easy identification. With the AKG 426B you can also rotate the complete microphone head through 45°, allowing quick changes from MS to AB techniques while the microphone is in its mount.

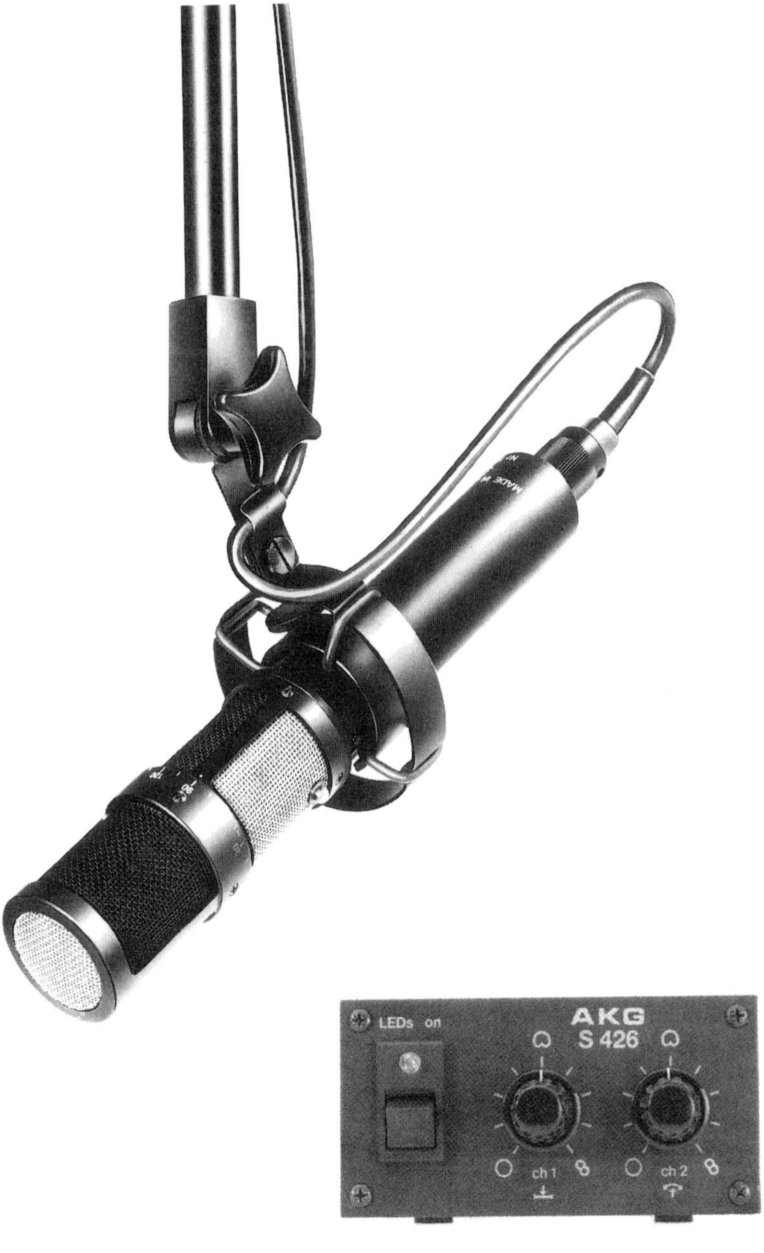


**Figure 3.21** One-piece side-fire microphone with remote control.

Microphones of this type are unlikely to become standard equipment on location for television, as they tend to be rather large and unwieldy, but for special, larger scale, set pieces, where you need a number of pairs, they are simple to rig and offer high quality sound. Most come with their own unique connectors, which plug into a multi-way cable and special extra boxes such as polar pattern selectors or matrices. The accessories all seem to be different, as do the pin configurations, let alone the colour coding. Carrying spare cables could save valuable time on location. An awful lot of these microphones need phantom power and it is by no means unheard of for only one capsule to switch on. This can be particularly painful if it is one side of a figure-of-eight, as you get an output that could be mistaken for stereo. So always double-check your microphones before you hang them 10 m above you.

## *Spaced AB microphone techniques*

In television mono compatibility will always be of the utmost importance. Coincident and near-coincident miking does, in most circumstances, give a good mono signal. It can, however, limit your use of the stereo effect. The recording of time-of-arrival differences, using spaced pairs, is potentially damaging to the mono but can substantially enhance the stereo effect.

With coincident techniques the centre of the sound stage is not on the main axis of the microphones. Although in practice this is not a major problem, moving to a spaced pair of either omni-directional or directional microphones that point directly at the action may possibly enhance the frequency response.

Another feature of coincident miking is that edge images appear lower in the mono mix than they do in the stereo mix. In many cases this can be a positive factor, as these are normally background atmospheres that are acceptable in stereo but distracting in mono. However there will be times when you will want these elements to be heard in the mono mix. Spot microphones are an option but these may well change the perspective. Moving to a spaced pair will minimize edge image drop-off without the need for extra microphones.

Simple spaced pairs, anything from near-coincident to around 10 m, receive both level and timing differences that are dependent on the distance and direction of the source. They give a sound stage that some people find more attractive than that produced by coincident microphones.

As with all stereo miking techniques, the placing of spaced pairs has a profound effect on the linearity of the stereo image and the subsequent mono mix. Although there are too many factors involved to predict accurately the mono compatibility, a sound stage that is basically symmetrical, and has a linear central cross-over, generally gives an acceptable mono signal.

### *Directional spaced pairs*

Using directional microphones, you have two options with the placing. You either cross the polar patterns or you don't. Crossing the pick-ups means that central sources will be on axis to the microphones so are reproduced accurately in both stereo and mono. Although the mono will have the same open feel as the stereo, there will be little of the mono 'tightening up' that is a feature of coincident miking.

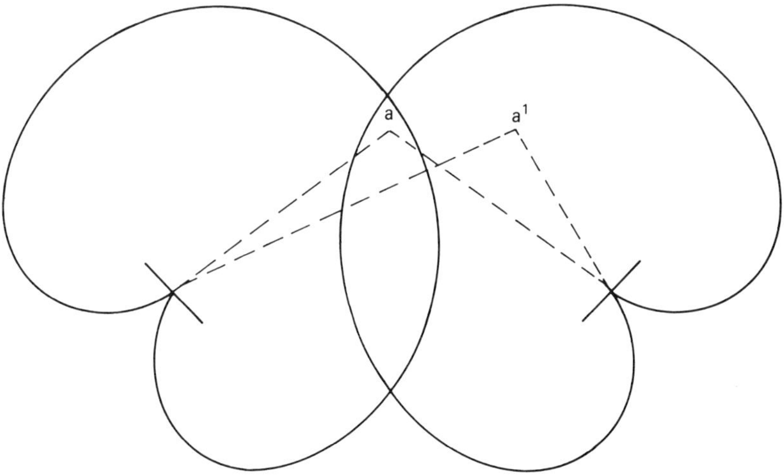


**Figure 3.22** Directional spaced pairs with patterns crossed. Out-of-phase effects occur on off-centre image at a′.

Sources that are off-centre will show the frequency-dependent cancellation in mono that is a function of the time-of-arrival differences. It is these very differences that can add some pleasing space to the stereo image. You will find, however, that the more directional the microphones, the greater the problems with mono compatibility.

If you elect to keep the microphones' pick-up as separate as possible this may reduce the effect of timing errors, but they will act more as spot microphones than as a pair, although if used in a very reverberant acoustic they can produce highly effective stereo.

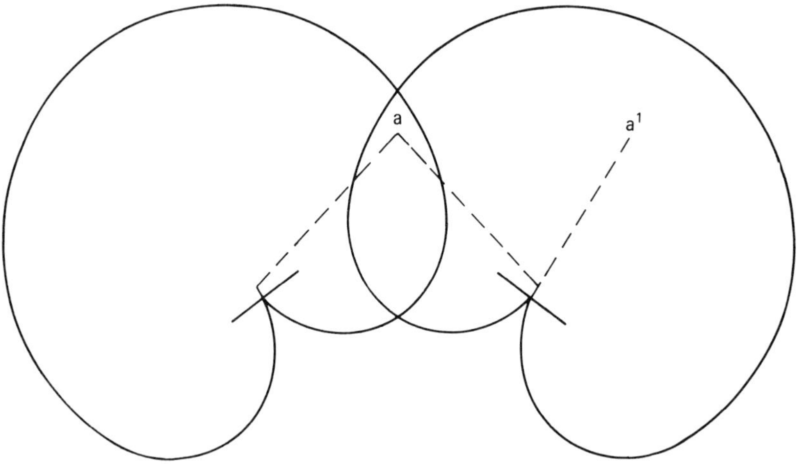


**Figure 3.23** Directional spaced pairs with patterns not crossed. No phasing occurs on image of off-centre source a′.

Any stereo image will be distorted if a spaced pair is placed close to the source, but it is worth mentioning, yet again, that we are not working in isolation. We are only part of the process and the pictures may well be so disjointed that the idea of a real sound stage is totally false. If that is so then there is no reason to limit yourself to a coincident pair that may well be further from the action than you would wish, just to obtain a stereo image that you don't need. Switching to a spaced pair, used in conjunction with pan pots (panoramic potentiometers), will allow closer miking and, with practice, some very acceptable stereo can be produced.

Because of the importance of mono compatibility, the use of spaced microphones as the main pair may be limited; but they can be used as a secondary pair with a main coincident pair or mono shotgun to 'open up' the sound. They are especially effective at the backs of rooms or halls picking up the reverberation or room acoustic. It is even possible to control the amount of reverberation that arrives in the mono mix by subtly panning the spaced pair.

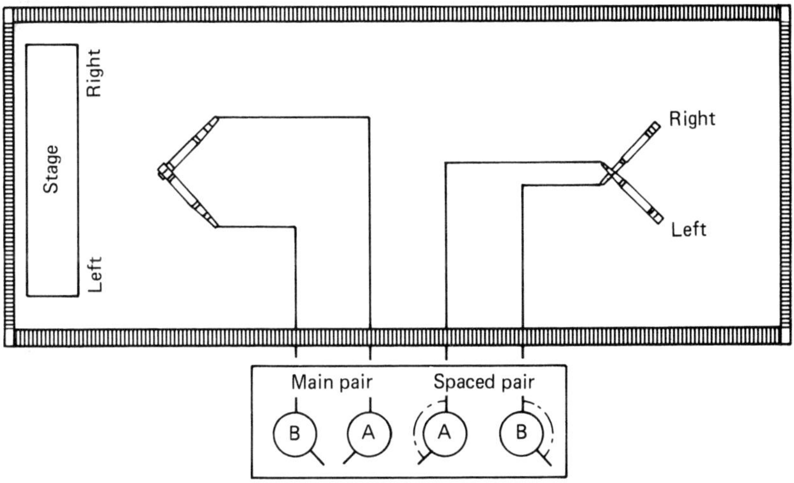


**Figure 3.24** Pan spaced/near coincident pair towards the centre to increase reverberation in the mono mix.

If you narrow the spaced pair by panning them more centrally you increase the direct-to-reverberation mix for the mono listener. This is a useful technique but should not be overdone as the reverberation must never become a point source in the stereo mix.

## *Spaced omnis*

In the music recording industry omni-directional microphones are favoured by many because of their accurate and predictable frequency response. So it is not surprising that methods have been developed for their use in stereo for television. Coincident techniques rely on directional characteristics so are pointless with omnis.

Spaced omnis produce a stereo image but do tend to suffer from a hole-in-the-middle. This led to an ingenious system pioneered by Decca called the Decca Tree. It uses three spaced omnis configured as a spaced pair with a closer central microphone. As one might expect, mono compatibility can be a problem but if used in a generous acoustic this does not appear to be the case. Music recordings made in this way offer refined stereo and well-balanced mono.

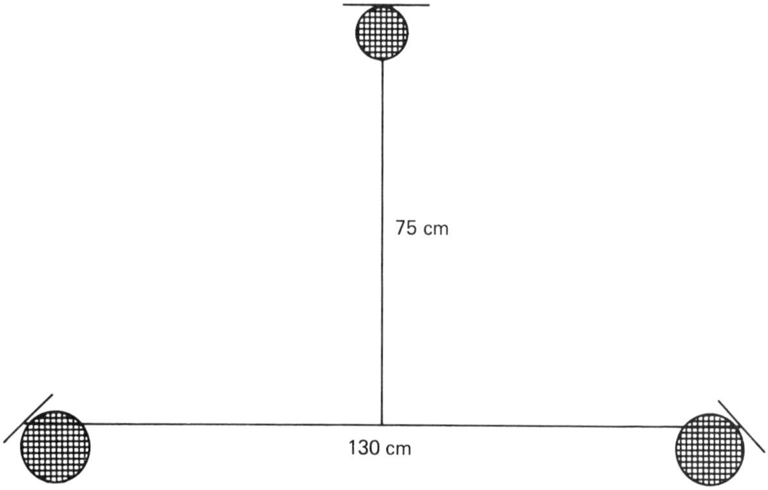


**Figure 3.25** Decca tree – three omnidirectional microphones.

## *Pressure zone microphones: PZMs*

In recent years a new form of microphone has emerged – the pressure zone or boundary-effect microphone. These use a rear-facing omni-directional capsule mounted no more than 1 mm above a back plate. Normally fixed to a surface, the microphone receives no reflected sound from the surface it is mounted on. Therefore any comb-filtering that might have occurred is eliminated. The result is a hemispherical response that is surprisingly flat across the whole audio frequency spectrum. Used as spaced pairs they can give excellent results in stereo.

Crown, a manufacturer of PZMs, have developed a 'Stereo Boundary Mount' (Figure 3.27) using either two PZMs or B&K Omni-directionals in a near-coincident array. The mount is designed to shape the pick-up of the microphones to minimize the overlap at high frequencies, thereby reducing mono phasing, while retaining much of the stereo ambience of near-coincident miking.

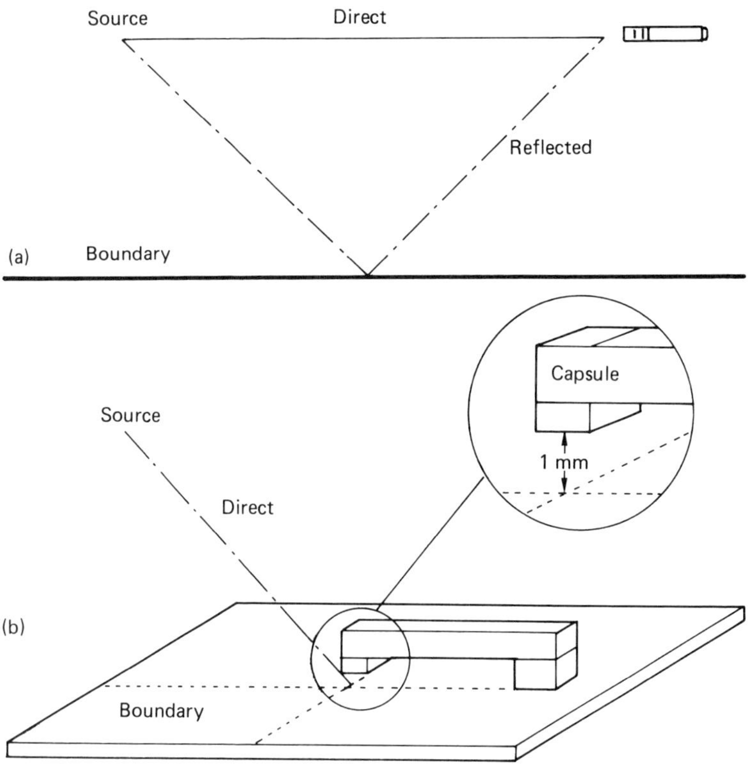


**Figure 3.26** Reflected sound arrives after direct causing comb filtering (a); direct and reflected waves effectively in phase (b).

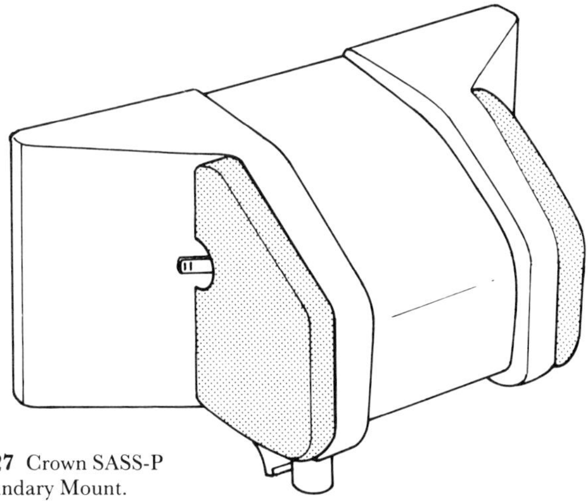

**Figure 3.27** Crown SASS-P Stereo Boundary Mount.

## *Binaural*

There are other forms of spaced microphone techniques. Binaural recordings are made using two spaced omni-directional microphones mounted either side of a dummy head. These are then reproduced on headphones. Many people find the stereo very convincing but find that the image appears spread out behind them, rather than in front. Binaural recordings are not reproduced well on loudspeakers and suffer from the mono compatibility problems you would expect from recording time-of-arrival differences. Their use in television is therefore expected to be limited.

## *Moving stereo pairs*

Television microphone techniques have always relied upon the ability of the microphone to move with the action. This presents us with major problems working in stereo. For, although we can accept quite drastic picture cuts within a scene, experience has shown us that we expect our sound stage to remain fairly constant.

Let us relate this to our experiences in life. Our eye/brain combination, like a camera, has the capability to zoom-in to a subject, without us having to physically move our bodies. We visually focus on individual elements of a scene and although the 'cocktail party' effect allows us to discriminate between sounds by direction rather than level, we do this in a very mobile sound stage. Rapid head movements shift our sound field dramatically, but if we offer this to the viewer, it is extremely disconcerting.

It seems that we use our balance sensors in conjunction with our hearing to smooth out the movement. Without this inertial navigation system, changes in the sound stage cause symptoms reminiscent of motion sickness.

If we work on the premise that wild changes should be avoided, we are left with the problem of how to cover a scene without individually miking every source. This would not only be impractical in many situations but would take away any perspective control from the operator. Therefore we simply must be able to follow the action using the minimum number of microphones.

Planning microphone moves through a stereo sound stage is not as daunting as it may first appear. Full left or right, and up and down, forward or backward moves, are still possible, depending on your distance from the source. There is only one movement that is to be avoided and that is left or right ROTATION.

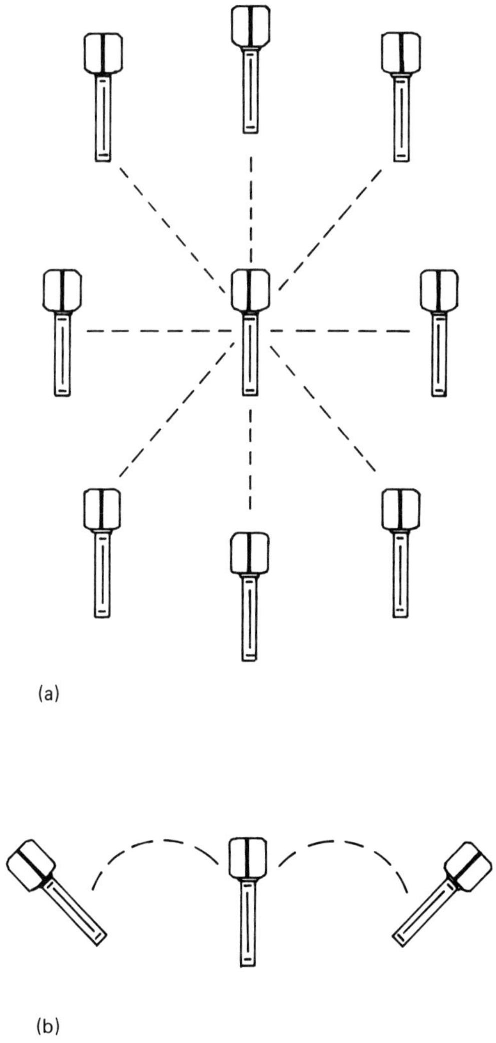


**Figure 3.28** Possible moves with a stereo mic (a). Rotation (b) should be avoided.

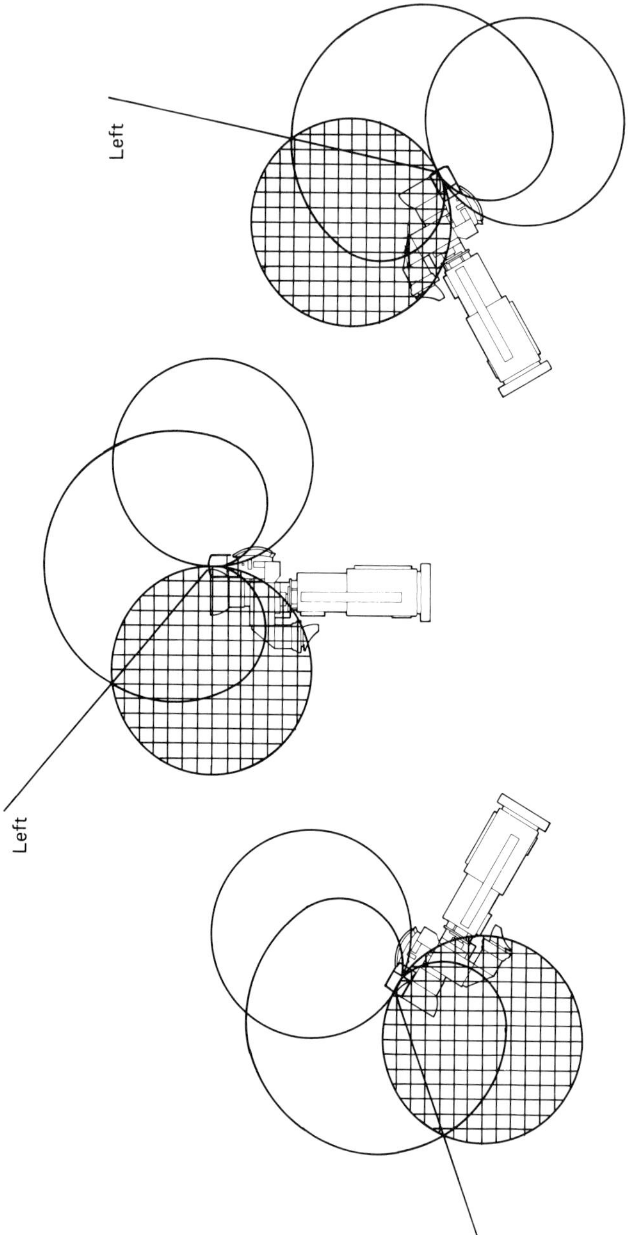


**Figure 3.29** Stereo mics on camera. Although the stereo image matches the camera shot, shifting background images are annoying to the listener.

## *Camera microphones*

We have seen that stereo microphones do have limitation on their movement, with rotation being the main thing to avoid. If you put a stereo microphone on a camera then rotation of the sound stage will occur with each pan of the camera. This is not acceptable and, regardless of the fact it matches the shot, it is terrible to listen to. Also the figure-of-eight element is likely to pick up noise from the camera and the operator.

## *Where do you point a stereo pair?*

It is relatively simple to aim a mono microphone although, as with all location recordings, the off-mic positioning is probably as important as the on-axis action. These same parameters control the crossed-pair microphone but, in practice, deciding what is on- and off-axis can be confusing.

You can simplify the whole thing by going back to basics. Regardless of their actual configuration, think of the stereo microphones' patterns as a separate middle (M) and a side (S) microphone. This will allow you to approach the problems of each individually.

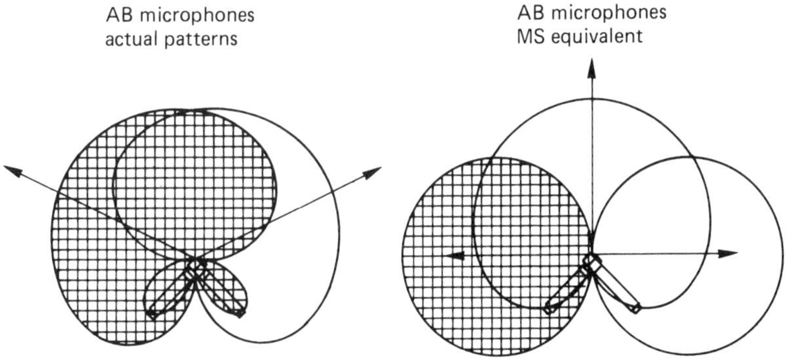


**Figure 3.30** The planning of moves is simplified by thinking in MS even when using AB configurations.

The M microphone you point at the action: that is relatively obvious. The difficulty is in keeping the S microphone parallel to the action. This normally means a lot more physical movement by the pole operator. One could imagine the microphone as working like the needle of a compass that keeps pointing North regardless of where you hold the case. There are people working on universal-type joints for microphone mounts but none is available at present.

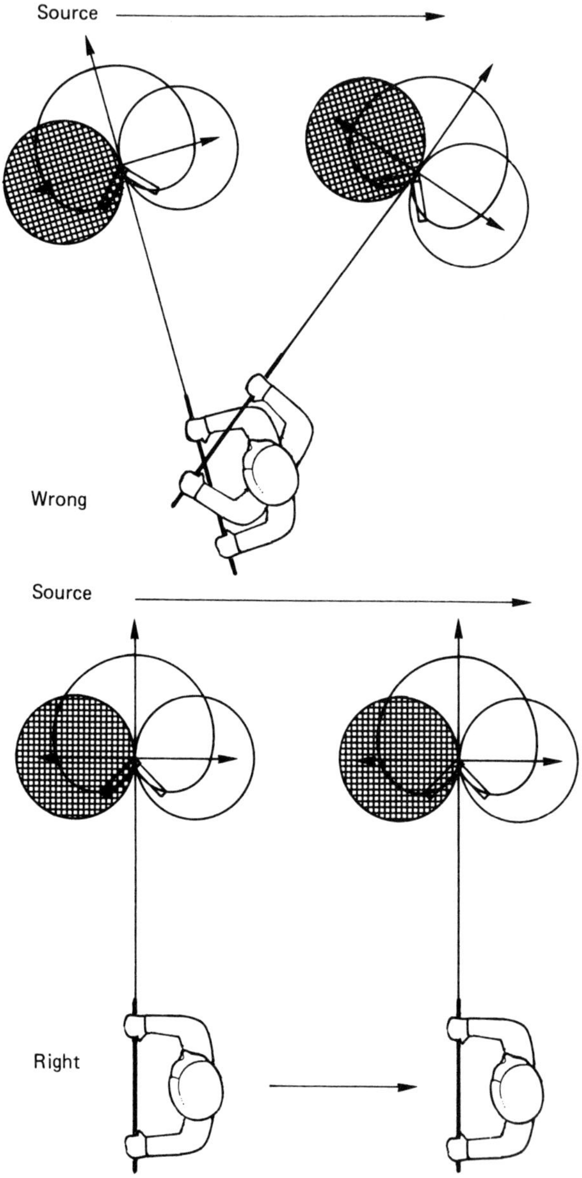


**Figure 3.31** Moving a stereo microphone. Keeping the mic parallel to the action avoids background shifts.

## *The limits of mobile AB pairs*

The new developments in microphone manufacture do allow the operator to contemplate moving an AB pair but aiming remains a major problem. Great care is needed but this should not be beyond the capabilities of a skilled operator.

Using our standard crossed pairs, we will probably have to accept a somewhat looser sound in mono than we are used to. This may not be a bad thing. In some instances, we have perhaps become obsessed with purity above realism, but there are many more occasions where a highly directional microphone gives directors more options with their shots. They will rarely compromise these for the sound. This is where the limitations of a single stereo pair begin to show. In many cases it is simply not possible to get the pair close enough to the action. It is not that it produces bad stereo, because it certainly does not; but it does give us major problems with mono pick-up. One solution would be to mix a shotgun or personal microphone into the stereo sound stage provided by a main pair, but this produces other problems we will discuss later.

As mentioned earlier, you have control of the width of AB pairs by the simple expedient of physically changing the angle of the microphones. But remember, regardless of your technique, an MS converter will give you remote and active width control while using an AB pair.

## *Using two stereo fishpoles*

There is a limit to what you can cover with one microphone, or pair.

Although every microphone you add will in some way dilute the purity of the stereo image, logic dictates that level, quality and intelligibility should have a greater priority than the stereo image. If the mono pick-up on a single stereo microphone is not adequate and if, in the same situation in mono, you would have used two microphones, then it is probably possible to use two stereo microphones. Never fall into the trap of using the lobes of your S pick-up to increase your coverage. In mono this area is off-mic: quite simply, it does not exist.

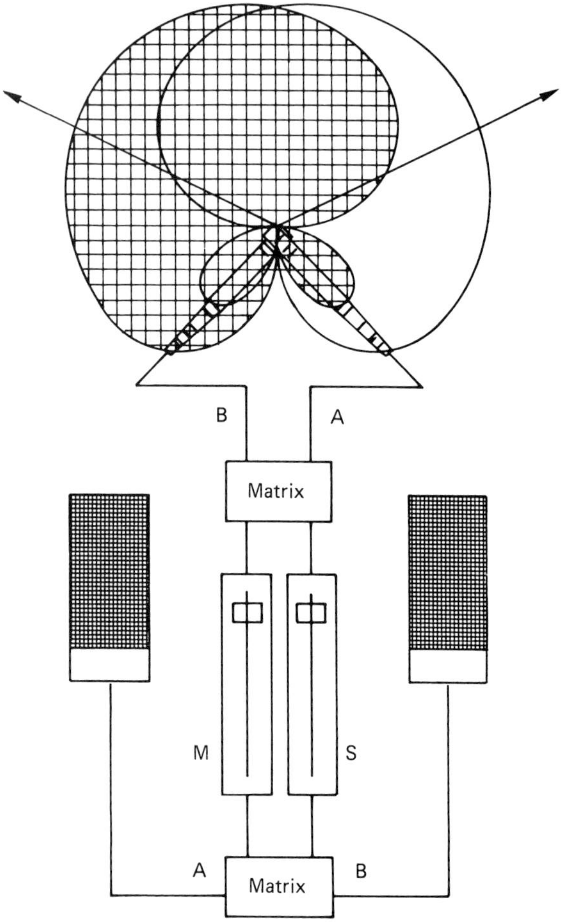


**Figure 3.32** MS mixing with AB microphone techniques.

Having mounted your microphones on their respective poles, go through the same procedures as you would for a single pair:

1. What is in the shot? Are there any sync sound sources other than the dialogue?
2. How close can you get your microphones to their main sources, taking perspective into consideration and without getting them or their shadows in shot?

3. What is to be off-mic? Use this information to angle the off-axis of both your equivalent M and S patterns. The opposing pairs' sound stage should be considered a 'no-go' area. Remember, the background sound stage should be the same on both pairs and remain constant throughout the scene.
4. Does this sound stage work with the pictures? As we have already seen, that does not mean it should necessarily follow the camera's view.

The range of movement of the individual fishpoles is the same as any stereo mic, so rotation is the only action to be avoided. The added complication is sources crossing opposing lobes of the pairs. Try as much as possible to work with the lobes parallel.

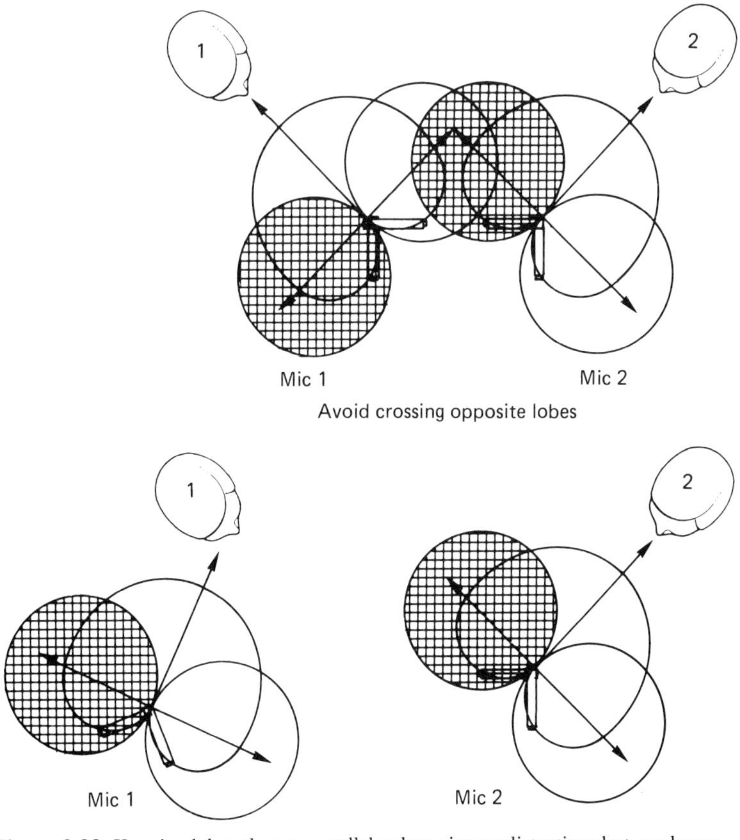


**Figure 3.33** Keeping lobes close to parallel reduces image distortions but can lower separation. Compromise is the best option.

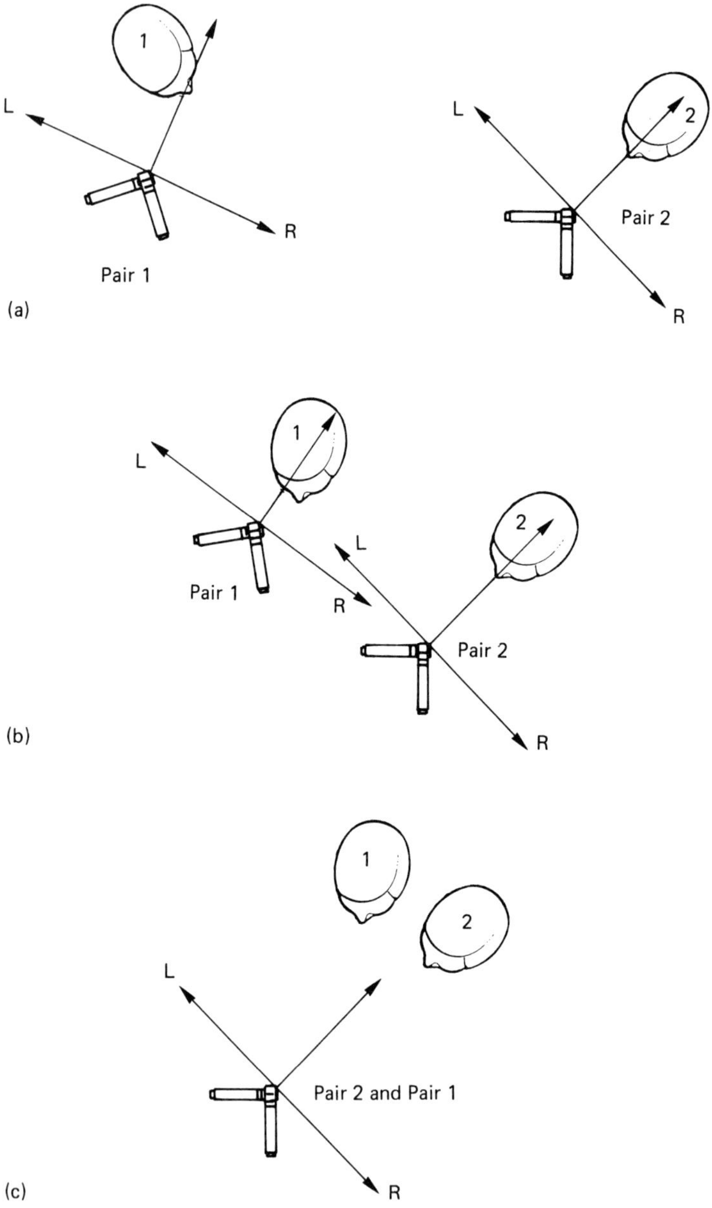


**Figure 3.34** A compromise between keeping the lobes parallel and the sources separate (a). The Haas effect will cause the image position of the closer microphone to dominate (b). When an overlap becomes inevitable, move pairs to a coincident point (c).

If you keep the pairs off-axis from each other, and have both faded up, there will be no vast image shift as they track with the action. Although central images on one microphone may well be to the side of the other microphone, the Haas effect will draw the listener to the image position dictated by the closer microphone. What you must try to avoid is sources moving from the lobe of one microphone to the opposite lobe of the other as this will cause unacceptable image shifts. When an overlap becomes inevitable, move the pairs to a coincident point as soon as possible, or adjust the lobes so that movement across them is minimized. This may mean some minor rotation of the pairs but this is preferable to central image wander.

If the background sounds are very diffuse you might consider reversing the lobes of one of the pairs. Then, although some off-centre movement would occur during cross-overs, it would be more natural than the centre-left-right-centre movement seen before.

As with all stereo microphones, think very carefully about the acceptance angle. If you opt for wide stereo, the image shifts across two pairs may become very distracting.

Cross-fades between pairs can cause image disturbances. Make sure that the main source is in the same position on each microphone before cross-fading. If you are working MS, dropping the S signal and making a mainly mono-to-mono fade is a good way of getting around the problem.

## *Mono spot microphones*

On location you will regularly find yourself in situations where the main stereo pair needs to be a fair distance away from the action for the acceptance angle to cover all the sources. It may then be necessary to spot-mic certain key sources, to control the balance adequately.

Obviously you would attempt to pan the mono microphone to the same place in the image as it appears on the stereo pair. However, any other source that is picked up by the spot microphone will also be pulled to that point. Therefore spill onto a badly-placed mono microphone will tend to narrow and distort the stereo image. The polar patterns and placing of all spot microphones should be carefully considered to keep unwanted pick-up to a minimum.

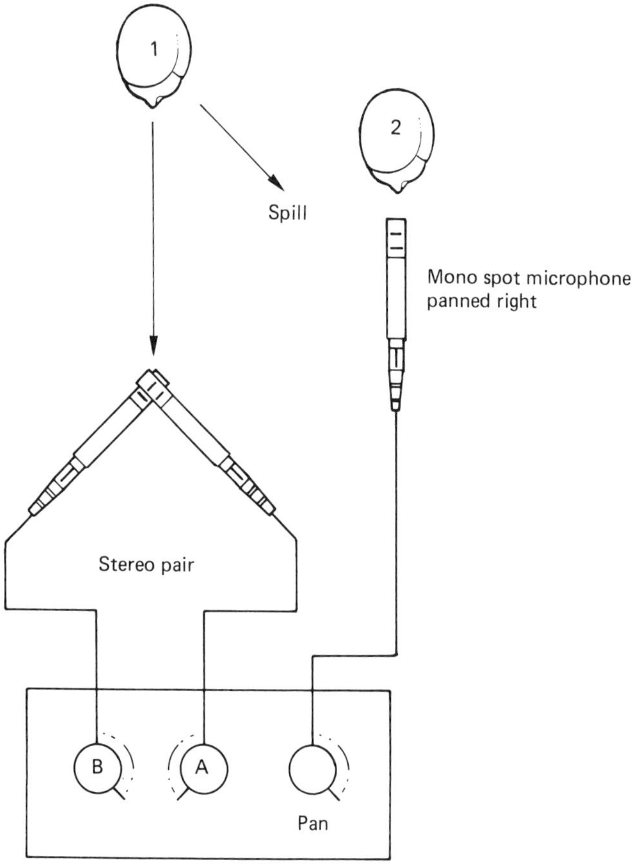


**Figure 3.35** Spill onto spot microphone pulls the central image to the right.

As the number of microphones increases you need to be aware of the effect of the timing difference between them. A mono spot microphone used with a covering stereo pair will hear a nearby sound long before the main pair. There will therefore be a timing/phase error between them. This will tend to dilute the stereo image as the phase coincidence of the pair is now compromised.

As with all phase discrepancies in AB stereo it is the mono listener who suffers most. Phase-related cancellation, comb-filtering, occurs as you fade-up the spot microphone, regardless of what polar pattern you use or how carefully you place the microphone. If two microphones are different distances from the same source, then a time delay will exist between them. Its effect will depend on the amount of delay and the level of each microphone in the mix.

Thanks to the drop in the price of digital delay devices, it is now possible to consider taking out the physical delay introduced by the microphone spacing. If you know the distance separating the spot microphone from the pair, it is possible to calculate the delay. As a rough guide 1 m is equal to about 3 msec. If you delay the spot microphone's signal so that it arrives at the same time in the mix as the pair, then the comb-filtering is eliminated. Remember that spill from other sources onto the spot microphones will be delayed by the same amount and because these are further from the microphone, this delay will be incorrect, and again comb-filtering would occur with the main pair.

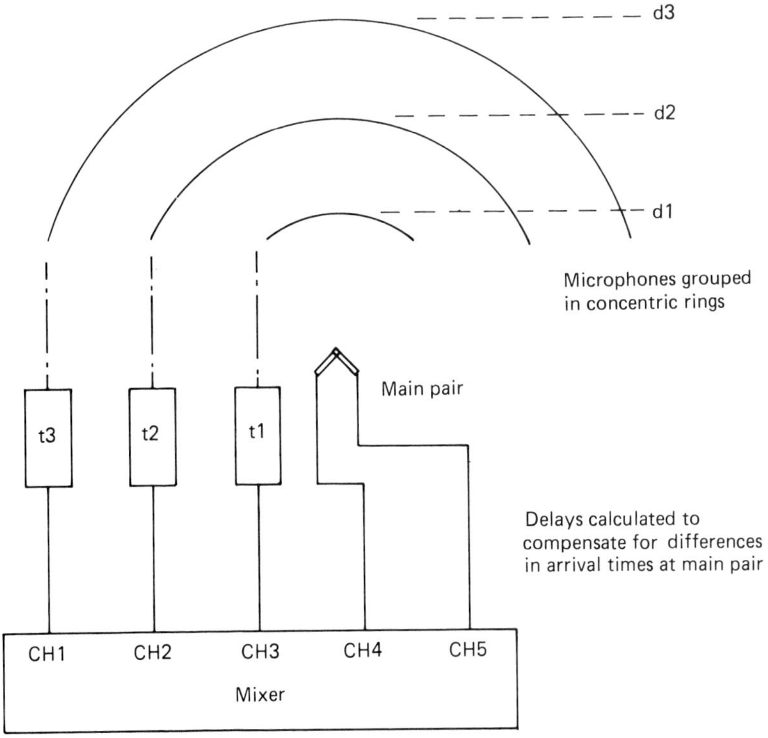


**Figure 3.36**

The nearer a sound source is to a microphone the better its frequency response. So a close spot microphone would theoretically offer better quality than a distant stereo pair. In practice this improvement is minimal because of the timing errors introduced. But using the delay technique mentioned above, many people find an improvement in the quality of the sound. Subjectively, some balancers feel that fading-up delayed spot microphones, to match a close-up on the screen, gives an improved frequency response that lifts the sound to enhance the pictures, without distorting the balance with the timing errors associated with non-delayed sources.

It is common for two spot microphones to be used to lift the edge images of a stereo pair. Be aware that the pair acts as a forward-facing microphone in mono, so the mono mix will therefore consist of three spot microphones. It is advisable to use the spot microphones close in and angle them slightly away from the centre, so as to minimize the comb-filtering on central sources.

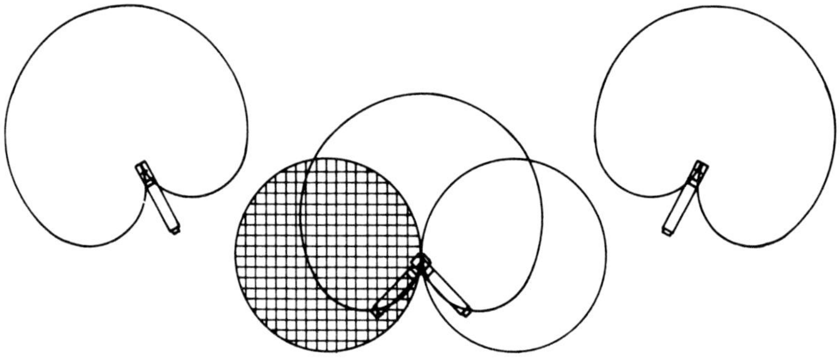

**Figure 3.37** Spot microphones angled to discriminate against central sources.

Balance spot microphones with great care. Remember that if you are monitoring in stereo you will not hear the comb-filtering that is taking place in the mono mix. Watch your meters as you change the balance. An MS meter will give you a good indication of the phase errors that you are introducing.

The more spot microphones you use, the more you corrupt the stereo integrity of the main pair. There comes a point when it is better to approach the whole thing as a multi-microphone mix and use the pair more for stereo reverberation than for imaging.

## *Stereo spot microphones*

As a camera moves through a scene the standard scenario consists of close-ups together with establishing and covering wide shots. In imaging terms this means that an element of the wide shot becomes the centre of attention in the close-ups. This element may be a single source which does not need a directional identity, beyond that suggested by the wide stereo pair. If it does then you might consider using a stereo spot microphone.

In its simplest form, a stereo spot microphone could be used to pull out a central source. The mix would suffer from all the problems associated with mono spot microphones; in addition it is quite possible that a source on the extreme left or right of the closer pair may well be effectively central on the wide pair. If the closer pair itself is off-centre of the main then these problems would be magnified.

The majority of stereo pairs have well-defined out-of-phase areas. If you use one as a close-in spot microphone, other sources may well appear in this out-of-phase region, thereby confusing the stereo image of the main pair.

You could narrow the image of the closer pair using the pan pots or the acceptance angle to minimize these positional problems, but you have to ask yourself, what is the point of using a stereo microphone if you end up panning it to mono?

Even inherently mono sources, e.g. speech, do benefit from stereo width, but it should be minimal. Although cross-fading between stereo microphones to match the perspective across a vision cut is not only practical but desirable, using a fixed stereo spot microphone in a mix is harder to justify.

## *Multi-microphone techniques*

There can be a tendency to leave the stereo image to the vagaries of the location, or the limitations of the placing of the main pair. We would strongly advise against this. Be sure to record what you intend to record. Sometimes circumstances dictate that the recordist needs total and separate control of a number of elements within a mix. This will often mean using a number of mono microphones rather than a stereo pair.

You might think that this makes for easy stereo but that is certainly not the case. You have all the problems associated with mono multi-miking, e.g. fixed perspective, phasing etc., together with the stereo problems of panning and image shifts.

The first step to good multi-mic stereo is separation. You only have full control if you can isolate the individual elements. On location this is never easy, so careful planning of microphone placing and polar patterns is essential. This 'patchwork' approach will inevitably lead to microphone 'dead spots' across the sound stage. Movement across these areas should, of course, be minimized.

The need for isolation does mean that you find yourself close-miking everything. This can destroy any realism, as perspective is a function of the distance from the source to the microphone. A microphone close to a source picks up the direct sound much sooner, and at a greater level than the reverberation. With a microphone far from the source, the direct sound and reverberation arrive closer together and will be of a more equal level. The high frequency response will also be diminished. It is these factors that give your perspective cues, and if ignored they will lead to major mis-matches between pictures and sound.

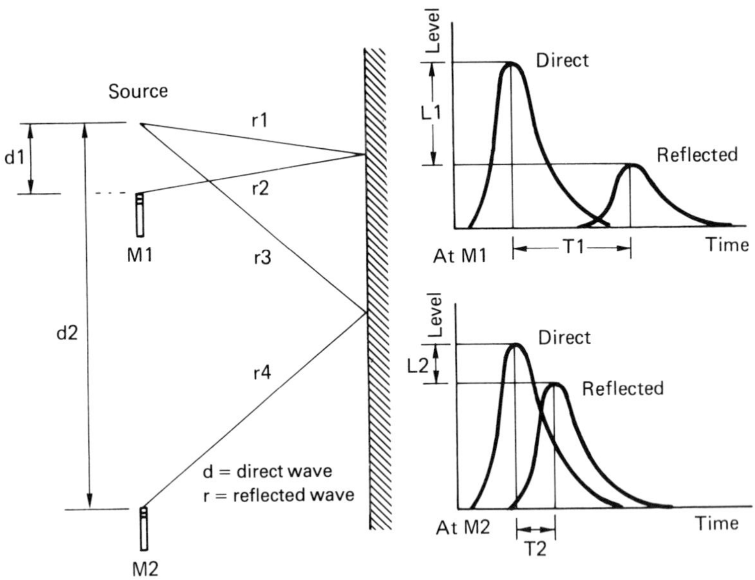


**Figure 3.38** The different set of values for level (L) and time (T), and their changing relationship gives powerful positional clues to the brain.

Accurate perspective control is extremely difficult using multi-mic techniques. Stereo highlights these problems to a far higher degree than mono. The very feature of increased ambience, which makes it worth while recording in stereo, now works against you.

It is possible to smooth the sound stage dead spots, and gain minimal perspective control, by adding a distant stereo microphone to your multi-mic mix. You must avoid any unwanted directional information being picked up, even to the point of aiming the pair directly at a close-reflecting surface, e.g. the rear wall. Otherwise its directional cues may conflict with, or restrict, your panning of the separate microphones.

Many multi-mic situations in television use radio microphones. These are almost invariably omni-directional and mobile. Separation is difficult and forever changing. Each microphone will pick up sound intended for the other. If you attempt to use a large sound stage then each will pull the others to its position in the image. This image-wander dilutes and confuses the stereo sound stage. However, if you mix so that the absolute minimum number of microphones are faded up at any one time and learn to be rather dynamic with the faders, acceptable stereo is a possibility.

## *Use of pan pots in a multi-mic mix*

All pan pots were not created equal. Different desk designers tackle the problems of creating a sound stage from different perspectives. From an operator's point of view, the particular type of pan pot law a manufacturer may have chosen might seem irrelevant. Unfortunately, unless you have total confidence that you can accurately judge the full sound stage and its mono compatibility on headphones, you should take the time to check what type of pan pot lies before you.

Regardless of the electronics employed, the physics dictate that it is impossible for both the stereo and the mono mixes to be unaffected by the position of the source. You need to know what compromises the designer has made. Panning may affect the stereo, leaving the mono relatively unscathed, or *vice versa*. More often you will find something between the two extremes. In this case, panning a source from the centre to fully left or right will lift the level by around 3 dB in the stereo mix, while the mono listener will hear a small decrease in level (see Figure 1.5, pages 12 and 13).

Because the 'cocktail party' effect helps you hear sources that are separated in space, you can find that the level increase with pan-potted edge images means you lower their level in the stereo mix. This, combined with the built-in level drop that the mono signal has already experienced, can lead to an over-emphasis on central images in the mono mix.

Having the ability to lay images across the sound stage encourages you to fill it. This is fine for music but great restraint is needed when recording multi-miked drama. All the problems of video editing in stereo are still there. Anyone who positions a source more than 20° from the centre line needs to know exactly how the piece is to be edited to avoid image shifts.

Actively panning the microphones to follow the picture is both physically difficult and annoying to the listener. Television screens are simply not large enough, yet, to match the available sound stage, and even if they were, the instant you edit the pictures, the stage is compressed or expanded to such an extent that the resulting image anomalies would be unacceptable.

In most cases you will find yourself having to multi-mic to narrow the image, rather than to widen it. Ask yourself, 'does this source really need to be off-centre?' Rarely is it sensible to use the sound stage 'because it's there'. It is worth remembering that the pan pot is a very powerful control. Resist the temptation to use it!

# 4

# Dynamic situations in AB

## Practical situations in AB

Let us examine a few standard situations and put a number of the configurations we have discussed to the test.

### *Situation 1*

One person in mid-shot directly to camera in a quiet rural setting.

#### *Version 1: crossed cardioids on a fishpole*

The resulting stereo is likely to be acceptable as long as you can work very close, somewhere in the region of 30–60 cm at most. This closeness will accentuate any head movements if you use a narrow acceptance angle, so it is advisable to loosen off the angle of the microphones.

This does mean that your choice of width is, to some extent, dictated by the position of the pair, rather than the ambient sound. Remember that if you wish to open this out with a separate microphone at a distance, a mono microphone will not be sufficient! Other microphones used for atmosphere must be stereo, otherwise they will appear as point sources in the image.

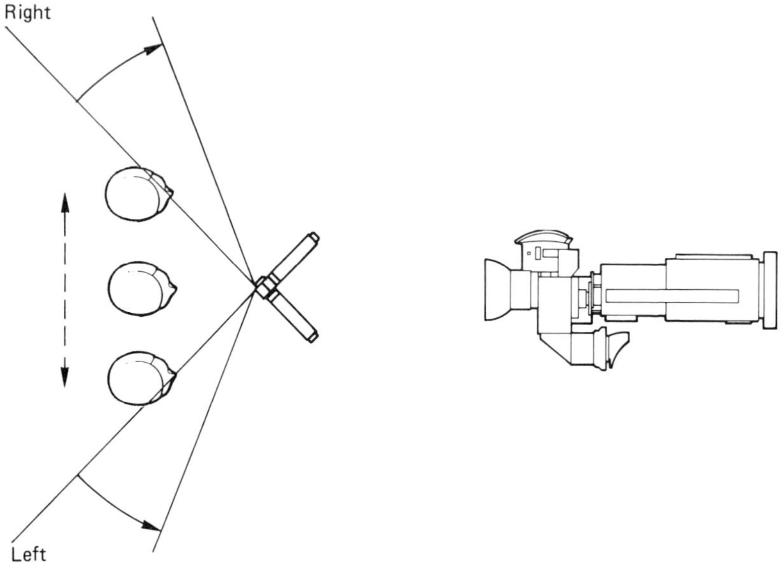


**Figure 4.1** Close working accentuates head movements. Use wide acceptance angles to minimize the problem.

The major drawback with crossed cardioids is their wide mono pick-up. Unless the conditions are almost ideal you will find them difficult to justify for television sound.

### *Version 2: crossed hypercardioids on a fishpole*

This will give you a little more scope with placing, but still needs to be relatively close. You might, under good conditions, find it acceptable out to about 90 cm. Setting the acceptance angle should be done with care as hypercardioids are prone to hole-in-the-middle effects.

One major difficulty with AB miking, as we have already discussed, is aiming, especially with side-fire AB microphones. There is also another problem that may seem obvious but is worth noting, and that is the physical weight of the unit. Two microphones on the end of even a medium pole will be extremely unwieldy and can become very wearing on the arms. This would probably be a good time to experiment with a detachable capsule system!

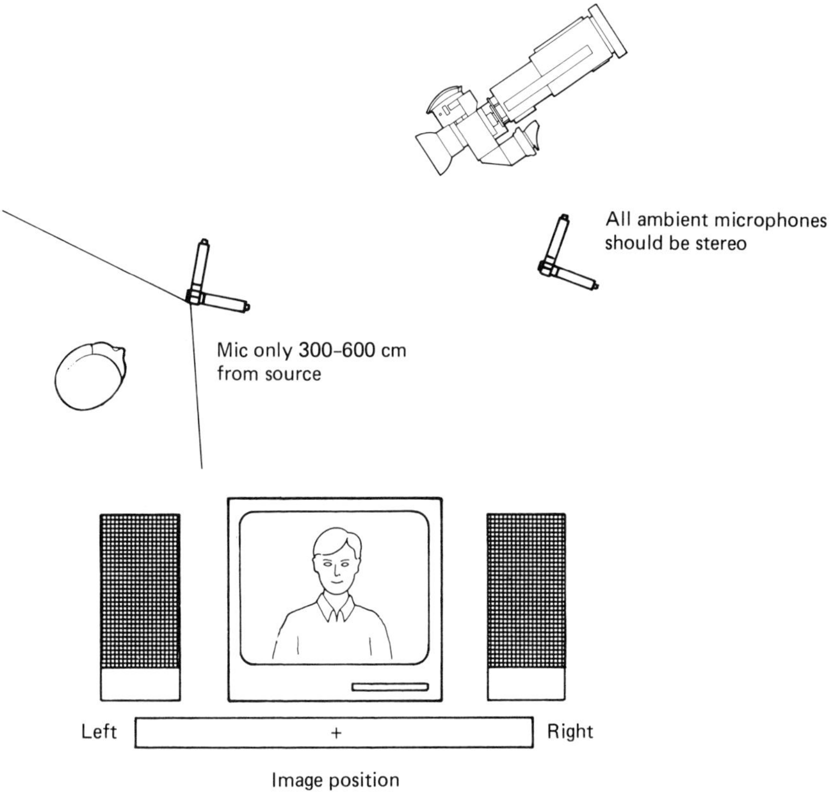


**Figure 4.2** Crossed cardioids. Use wide acceptance angles and work close in.

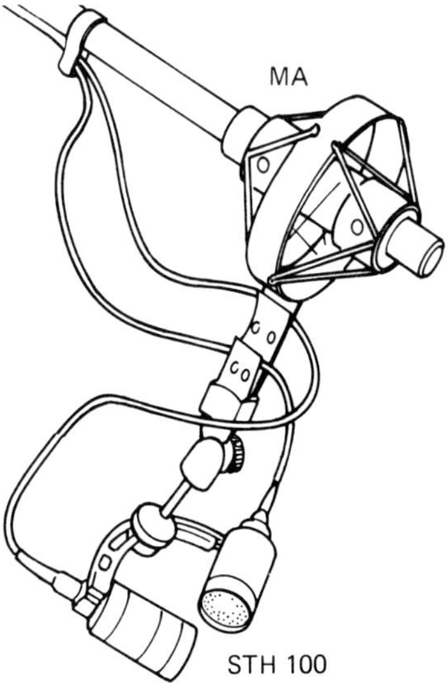


**Figure 4.3** AB miking using active capsule technology.

*Version 3: mono shotgun microphone plus a stereo pair away from, but on axis to, the action*

You can move the shotgun microphone without worrying about image shifts, and you will get a tighter sound from the shotgun than that available from any combination of stereo pairs. Although adding in the more distant pair for a stereo effect will open it out, this is normally not only acceptable, when listening in stereo, but may well be found desirable. However this technique has also opened out the mono, and it is sometimes very difficult to judge the two mixes' individual merits when switching from stereo to mono, as the stereo always 'sounds' better than the mono.

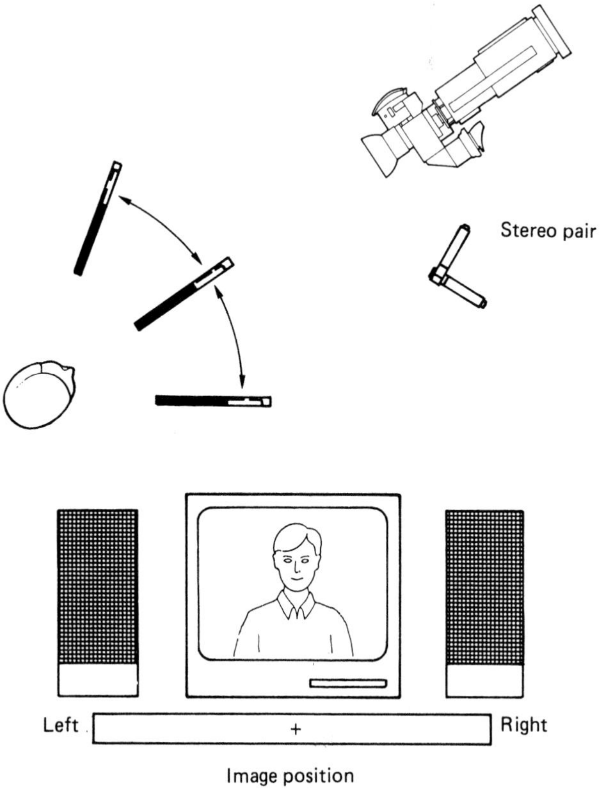


**Figure 4.4** Move shotgun plus stereo pair. You can move the shotgun without changing the image.

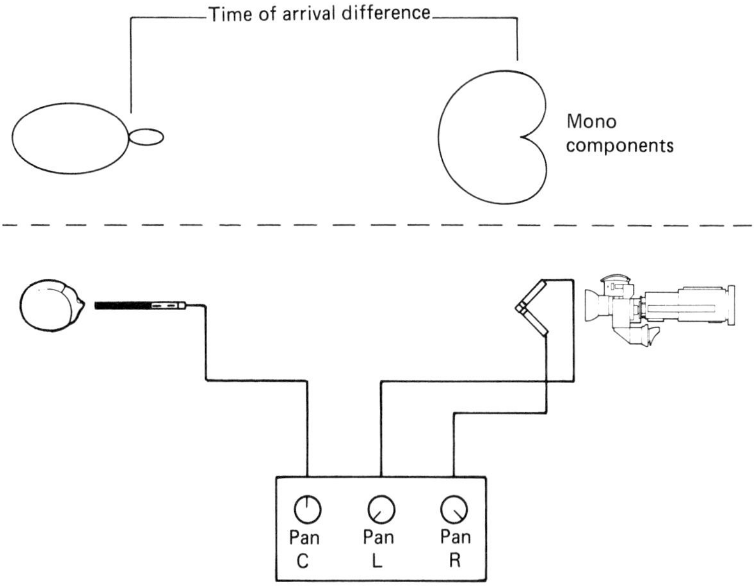


**Figure 4.5** Acceptable stereo, but phasing can occur in the mono mix on central images.

The resulting mix cannot be radically altered in post-production as there is no escaping the fact that you have recorded a time-of-arrival difference that cannot be reversed. Nevertheless the final product is stereo and don't overlook the fact that you can adjust your width by physically changing the angle of the microphones in the pair.

### *Situation 2*

One person to camera, starting in a wide shot, walking to camera and ending on a loose mid shot.

#### *Version 1: crossed cardioids on a pole*

The major problem here is simply pick-up. This configuration appears very distant in mono as it is equivalent to a 'fat' cardioid. If you can get close enough, and the background noise is low, they can be used to good effect but in most cases hypercardioids are easier to use.

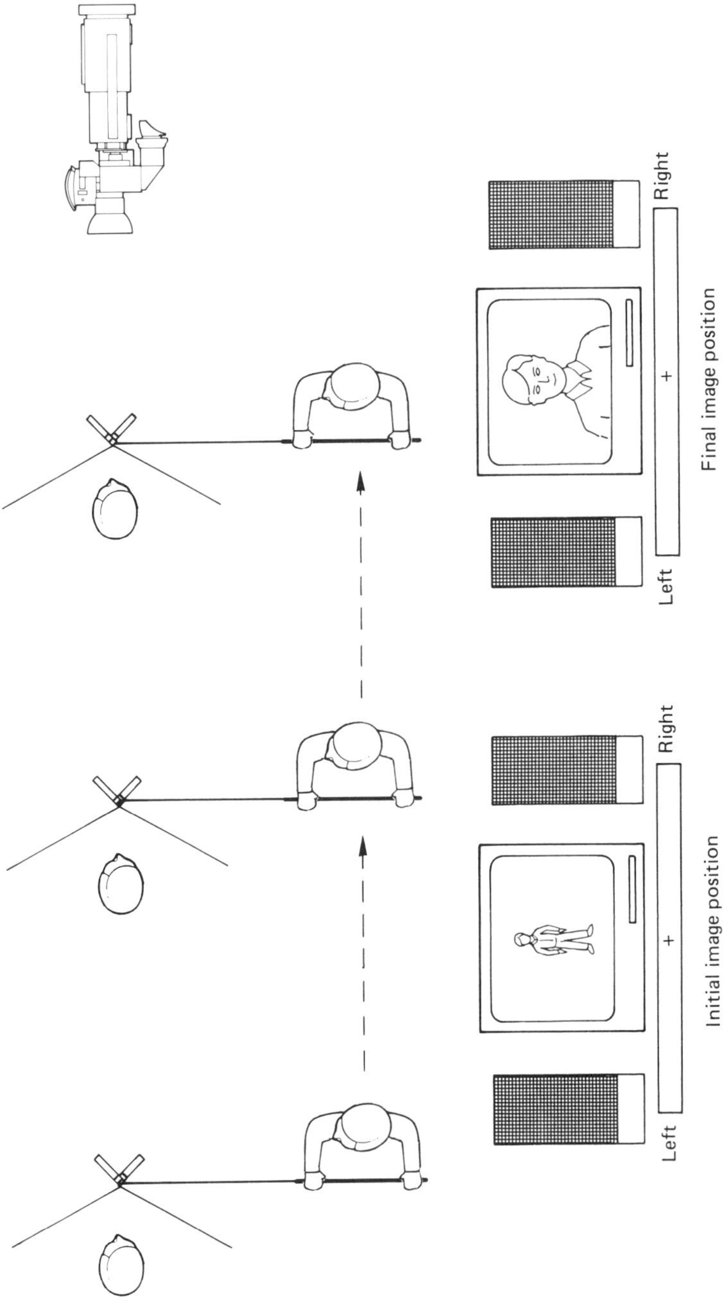


**Figure 4.6** Tracking with crossed cardioids or hypercardioids.

### *Version 2: crossed hypercardioids on a pole*

Let us look at the thorny problem of tracking the microphone. First, what is in the shot? Next, are there any sync sound sources other than the dialogue? That includes footsteps and camera noise.

These criteria give you your limits for placing. Be aware that although perspective can, and should, change, the background sound stage is set and must remain constant throughout the scene.

As mentioned earlier, it may help if you imagine your stereo pair as working as middle and side. So first you have the M component to position. The sound source is fairly obvious, but as any experienced recordist will tell you, the most important decision is what is to be off-mic. Therefore with a little thought you can make the front-back setting. Having completed that let us move on to left-right.

The fundamental mistake that is now made is to follow religiously the camera's view of the scene. Why? The S component is a figure-of-eight: just like any other microphone, put it where it hears what you want it to hear. In this case, if the background is very general, then it is possible to relate to the camera. However if, for instance, there was a tractor on a hillside off to one side, then try turning the figure-of-eight until the tractor was being picked up equally by both lobes. Few things are more irritating than background noises that are off to one side of the image. Remember that this MS approach to position is directly applicable to coincident AB miking.

The shot begins fairly wide and our pair, on edge-of-frame, matches the shot. What doesn't work is the well used sweeping-arc type of pole action, as this would change the sound stage. The pole has to move so as to remain parallel to the action. This does not preclude in-out or up-down moves. As the movement begins, the headroom decreases and just as in mono we follow it in.

Unless the shot gives a heavy off-centre bias on the dialogue, it is advisable to keep the image central, as there is a good chance that, if it is ever edited, the first cut will be to a close-up and it would be very disturbing if the sound image jumped at the edit.

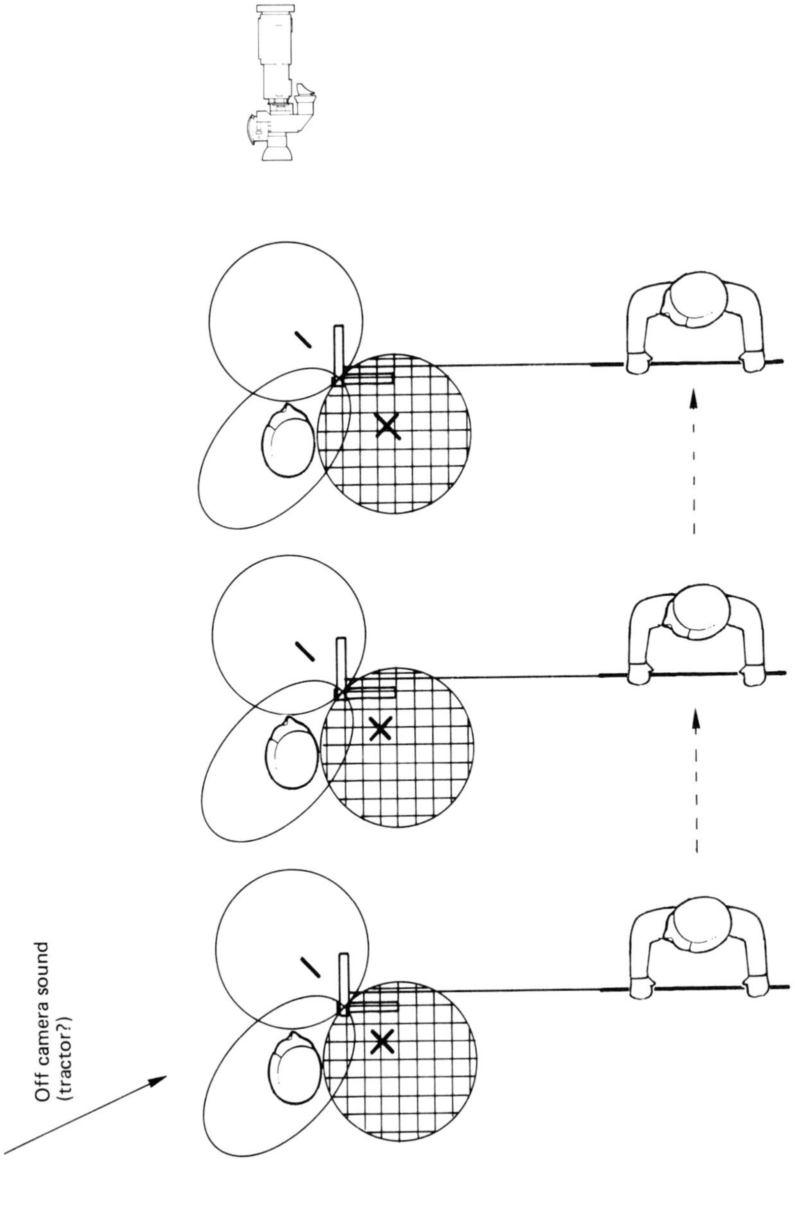


**Figure 4.7** Microphones angled to centralise background sounds.

*Version 3: fixed stereo pair with separate mono microphone*

In this instance the mono microphone is used exactly as it would be in mono. As with any microphones separated in space, we have a timing discrepancy to think about. Remember that the stereo pair not only gives you stereo width but also has a mono component. This will open out the mono mix, therefore the direction of this mono component is very important. You need to decide whether the pair is to be used to give position to the action or just stereo ambience. If you point the mono component at the action then it will have a greater effect on the long shot than if you pointed it away. But as the action approaches the pair the perspective will tighten considerably, as the timing error decreases, matching the shot exactly.

### Situation 3

Two people at a party standing in the kitchen; 2 shots, including over-the-shoulder, plus singles.

*Version 1: single stereo pair*

The normal first reaction to this set-up seems to be to maintain the left-right integrity of the main 2 shot, a single stereo pair between the couple giving us a 'true' stereo image. Sounds reasonable, but the resultant mono is not pointing at anyone! There is no way that the microphones can be physically panned, as in mono, as this would throw the stereo image all over the place. The only option is to move far enough away so that the mono pick-up includes the actors. And what if you need to favour a quiet speaker or, even worse, move in with the shot? This would actually increase the perceived width of the sound stage; precisely the opposite of what is required. This microphone technique fails in stereo almost as badly as it does in mono. 'True' stereo is sometimes just too difficult to work with.

Allowing one's camera shot to lead us into this wide-shot-equals-sound-stage approach is very common. It is absolutely essential that the sound recordist has a global view of each

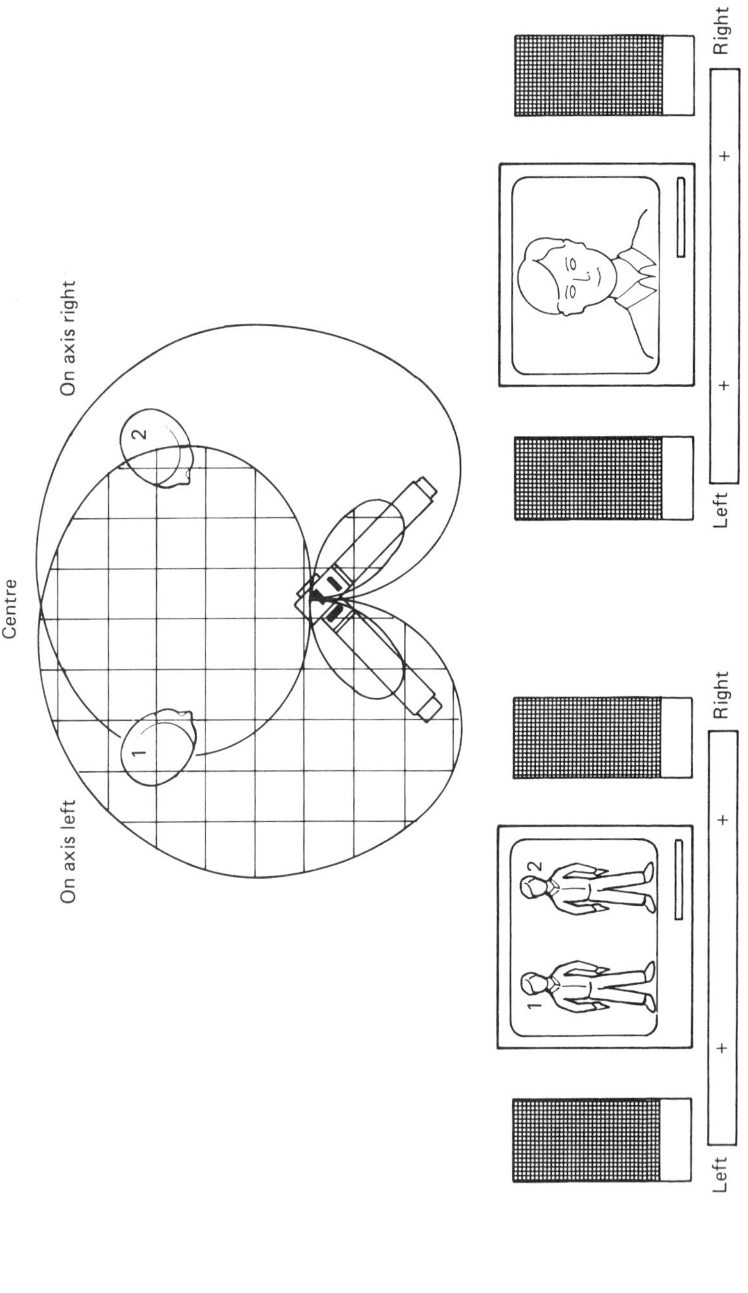


**Figure 4.8** Image position in wide shot (a): sound matches picture. Image position in close up (b): sound conflicts with picture.

set-up. If we place the actor who is on the left of the picture on the left of the sound stage, what happens to the image on the singles? We could easily end up with a tight close-up of an actor whose voice is coming from left of centre.

Sync sounds that have a real direction, e.g. a coffee percolator to the actors' left, may well be acceptable in the 2 shot but will become very disturbing as you move to the singles, as position now allows us to pick out sounds of a much lower level in a mix than is the case in mono. There are times when this is a positive advantage but these are rare.

We could, of course, rigidly mimic reality, matching our microphone placing to the changing camera shots. This is difficult if not impossible to live with on the screen, especially when you realize that these rapid shifts of image are happening to both participants.

### *Version 2: mono mic plus stereo pair*

It is necessary to look beyond what is real, and ask yourself what is required. Off-mic mono is not an acceptable by-product of stereo. So let's approach this scene again. What you want to record is the two actors in a stereo environment. If you think about it, the editor will work the scene so that the vast majority of it will take place with the actor who is speaking in the centre of the screen. Their true position will rapidly become irrelevant. So why tie the sound to this arbitrary criterion? It would simplify the whole process if the actors remained in the centre, but were surrounded by stereo ambience. Film makers do this by recording the dialogue in mono and dubbing stereo wildtracks plus, perhaps, some artificial stereo reverberation. Television could of course do the same, and in some instances will want to. However, there are a number of ways to achieve this effect on location thereby cutting down the dubbing time and producing a very real acoustic.

The simplest of these is to mono-mic the actors in some way, either with personals or a shotgun microphone, and have a stereo pair at a distance giving us the sound stage. This approach is beset with all the problems associated with recording time-of-arrival differences that we discussed earlier, in addition to the physical placing of the pair, if the kitchen in our example is small and full of a TV crew. It may well prove a fascinating challenge! It does, however, give us a full stereo recording in the can on the day.

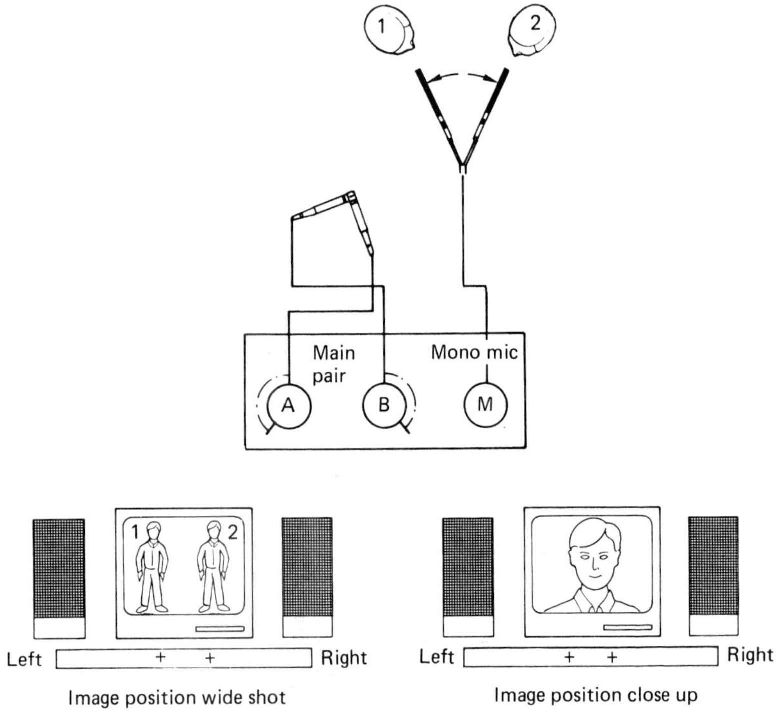


**Figure 4.9** Image matches picture in both wide shot and close up, while stereo ambience fills sound stage.

### *Version 3: separating the stereo image from the camera*

A better solution to this problem comes from going back to basics. Yet again let us remind ourselves that what is required here is central-ish dialogue within a stereo sound stage. We would prefer to use the minimum number of microphones, i.e. two, and if they could be coincident so much the better.

This is one of the times when it is helpful to think of the microphone as working in the MS configuration.

So first let us totally ignore the S and pretend we have a standard mono microphone. Where we place it is just as we always did – in this case probably over the top – and we follow the action. Perfect mono compatibility.

Now we tackle the S signal. Are there any sounds, other than the dialogue, that need a definite position? Answer 'no'. Therefore, provided nothing that we want in the centre appears on one or other lobes, the resulting ambience will fill the sound stage. It would therefore be acceptable to point the figure-of-eight parallel to the sound stage, with the actor on-mic, and move the microphone sideways to pick up the other actor's lines. This attempt to follow the action only succeeds if the microphone is always in position before the actor speaks, and the actors never speak at the same time. Otherwise image shifts would occur.

An elegant solution to this is to take our previous argument a stage further. The important criterion for this figure-of-eight microphone is its off-axis response. So why not turn it at right angles to the actors, i.e. front to back or even up-down? This technique is known as 'lollypopping' and is very useful in a whole host of situations. This decoupling of the stereo sound field and the camera's perspective is a very important step in your approach to location recordings.

### Situation 4

Two people at a table in a restaurant.

A single stereo microphone, with the camera's perspective, would be very difficult to work in this situation. Tracking of the microphone would naturally change the background sound, which is highly undesirable. The operator could possibly lollypop the microphone in some way but practically, using only one microphone, it would be very difficult.

Another approach would be the use of two stereo microphones on poles, one covering each of the actors. The problem is where to point their S lobes. If we match the camera, then what is central on one microphone is on the lobe of the other, causing image shift as we mix the microphones. Again the important factor here is the off-axis response. Front-back lollypopping is an

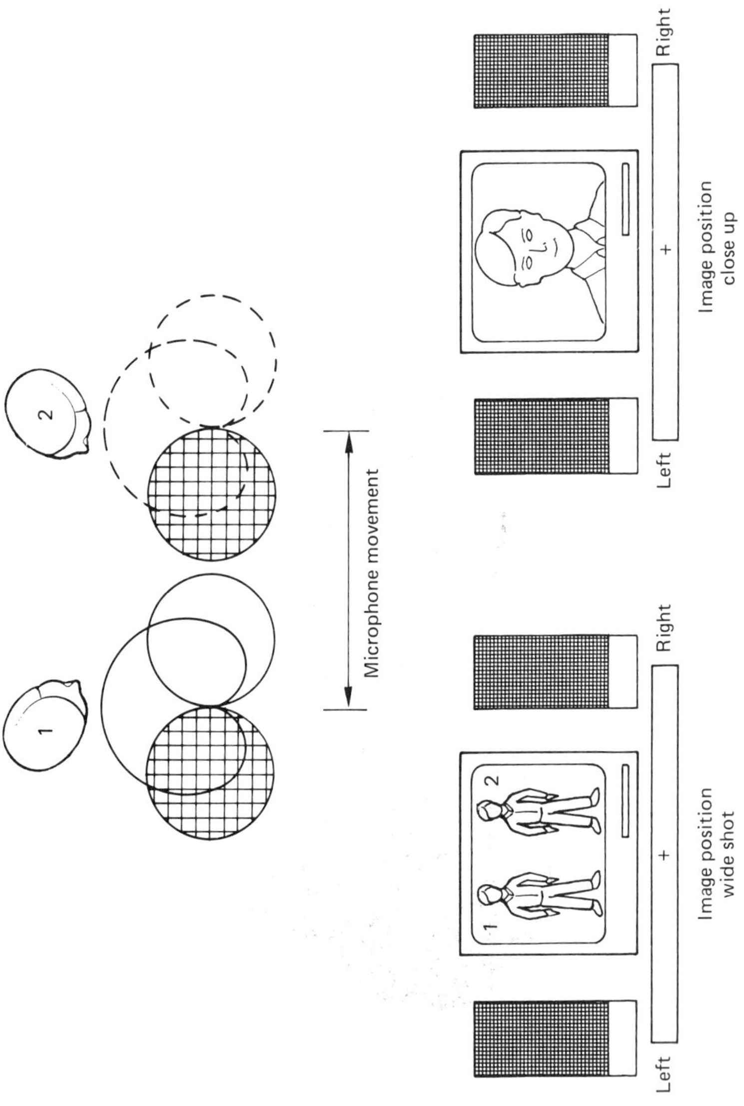


**Figure 4.10** Moving the microphone parallel to the action gives a stable image with a real stereo environment.

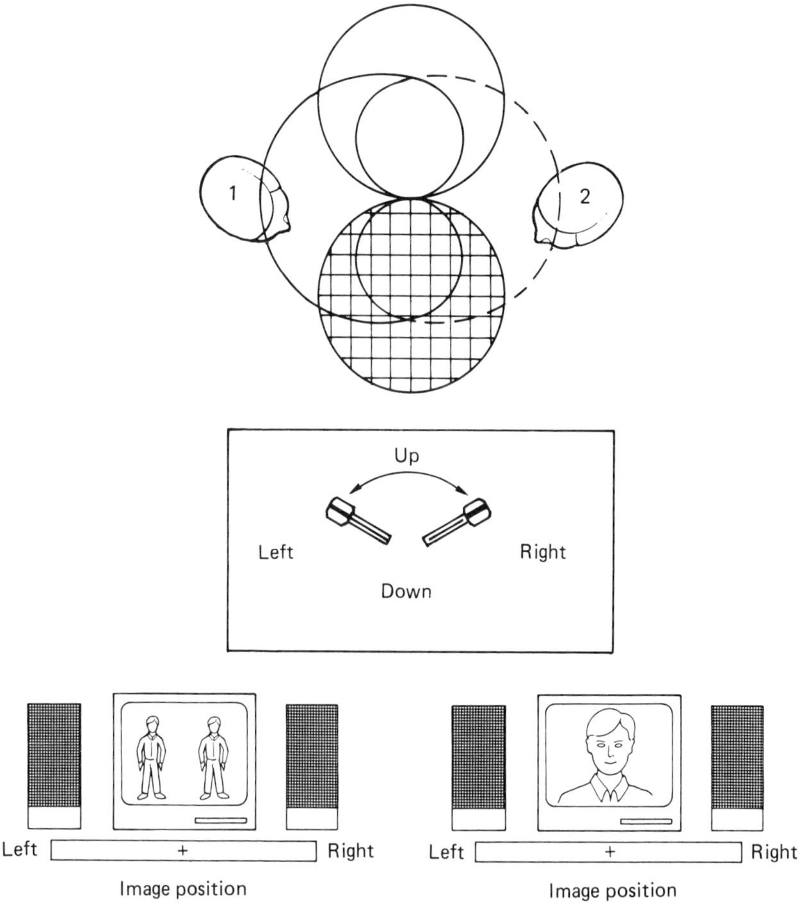


**Figure 4.11** The alignment of the S component of the microphone remains the same, while the M component moves with the dialogue.

option, with the positive lobes either pointing the same way or reversed, so as to balance the sound stage. Thus sound off to one side would appear on both positive and negative lobes, thereby smoothing the image. You may even find up-down could suffice in certain environments. What is important is that the dialogue remains central, otherwise editing will be a nightmare. Remember, this technique still gives you a full stereo environment to work in, only the dialogue is kept in the centre.

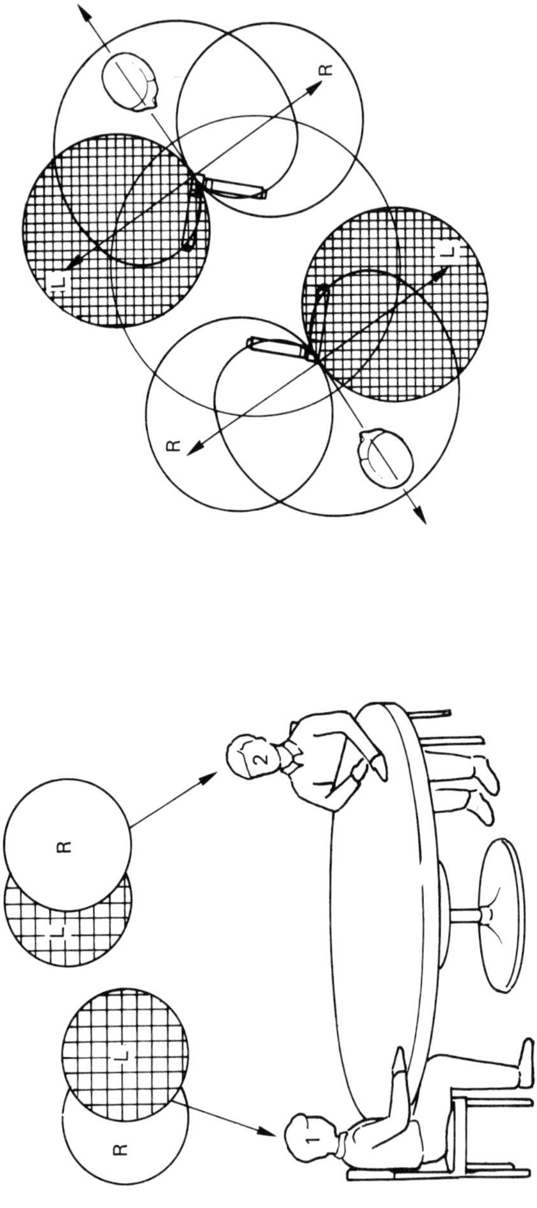


**Figure 4.12** Keeping the side lobes parallel and reversed will smooth the stereo background.

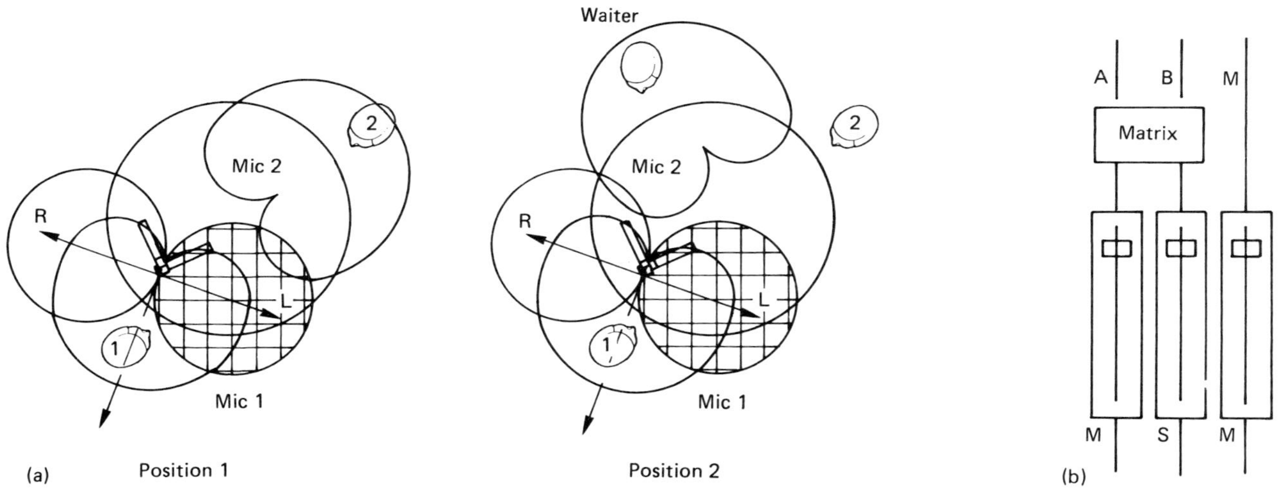


**Figure 4.13** Microphone 2 pans to cover both camera right subject and waiter (a); stereo background is provided by microphone 1. MS mixing (b) allows you to mix the mono sources independently from the S component.

### *Situation 5*

Expand the situation to include the head waiter arriving with the menu.

The problem now is in deciding which microphone should move to pick up the waiter. With two stereo microphones this can easily lead to image shifts. You can simplify the scene by switching to MS mixing. This will allow you to fade out the S signal from microphone 2, thereby freeing it to move between subject 2 and the waiter. You mono mix the two M signals whilst retaining the S signal from microphone 1 for stereo ambience. This method of mixing is explained further in Chapter 5.

#### *Handheld stereo microphones*

One can demonstrate all the horrors of badly conceived stereo using a handheld stereo microphone. But if you tightly control the parameters and the person using the microphone is aware of its limitations, then acceptable stereo is a possibility. A wide acceptance angle is essential; anything over 100° is OK.

For all the reasons already discussed, keeping the participants left and right is unworkable in most situations; therefore it makes sense to back/front lollypop the microphone. This keeps the background sound stage constant and the dialogue central.

# 5

# MS microphone techniques

## The benefits of sum and difference (MS)

These two letters seem to have caused more misunderstanding than any other aspect of stereo. Yet really it is a delightfully simple and elegant system. All you need to remember is some simple mathematics.

IF: $A + B = M$ (sum) and $A - B = S$ (difference)
THEN: $M + S = 2A$ (left) and $M - S = 2B$ (right).

Do not be intimidated by the minus signs. They simply mean phase reverse!

M, the SUM of A and B, is fairly simple to understand. The S signal, A−B, is a more difficult concept to grasp. Think of S as the DIFFERENCE signal, when interpreting meters, but use the SIDE concept when placing microphones. It is absolutely wrong to call the S signal the stereo signal, as a stereo signal must contain both M and S.

The mathematical relationship, shown above, can be related directly to microphones. If we consider a pair of crossed hypercardioids at 90° as our A and B signals, as we saw in Chapter 1, a plot of the sum of the polar patterns (A + B) gives us the equivalent of a forward-facing cardioid microphone, hence M the middle signal. Plotting the resultant of A − B, i.e. S, the difference, gives a figure-of-eight type response at 90° to the M. It is this physical relationship which brings about the side concept for S.

The beauty of this transformation is that it works in reverse. The sum of a forward-facing cardioid and a sideways-facing figure-of-eight is a hypercardioid facing left, while the resultant of M – S is a hypercardioid facing right.

Problems with aiming AB pairs in television booms was the reason for the adoption of MS microphone techniques because, as we have just seen, it gives you a mono microphone to point at the action. Please note we have said 'techniques', not 'microphones'; this was deliberate. We are somewhat disappointed that many operators see the new one-piece MS microphones as THE way to do stereo television. On the contrary, having separate microphones allows you to match the microphone to the situation in front of you, or perhaps to the left!

Taking the middle-side concept further, you can see that the relationship of S to M relates directly to the perceived width of the sound stage. Therefore altering the gain of S has the same effect as changing the acceptance angle of an AB pair. This dynamic manipulation of width is what first attracted television sound supervisors to MS mixing techniques, as it allows the matching of sound stage width to picture width.

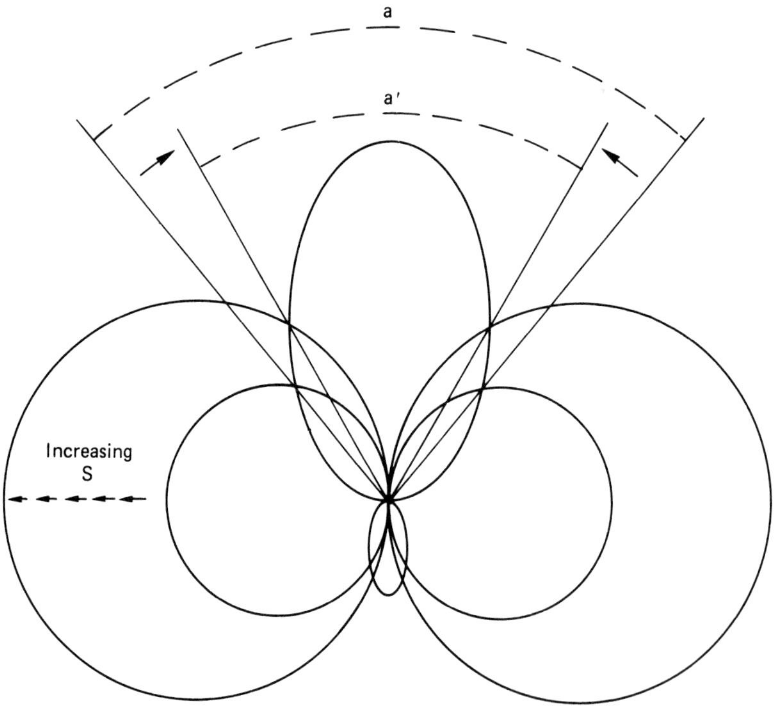


**Figure 5.1** Increasing the S gain narrows the acceptance angle a to a′, giving a wide stereo image.

So although MS microphones will have a large part to play, the real power of MS, for television, is as a mixing tool. Operators should look beyond the convenience of one-piece end-fire MS microphones, to the whole concept of width/ambience manipulation.

## AB and MS signal transformations

We have used the convention of plus and minus when transforming signals. However it should be pointed out that you cannot produce a combined signal by simply paralleling the source signals. If you do, the negative feedback of the output amplifiers fight each other, causing current limiting and therefore distortion. You need a virtual earth mixer to eliminate this problem.

(a)

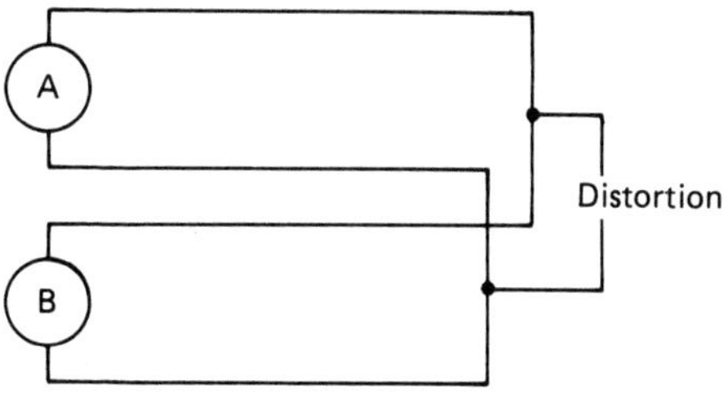


(b)

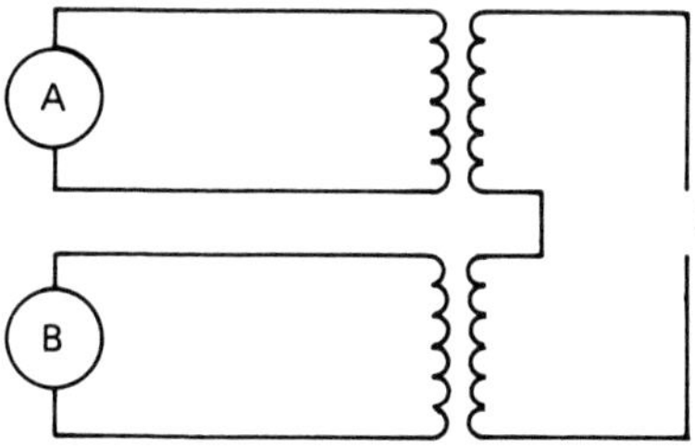


(c)

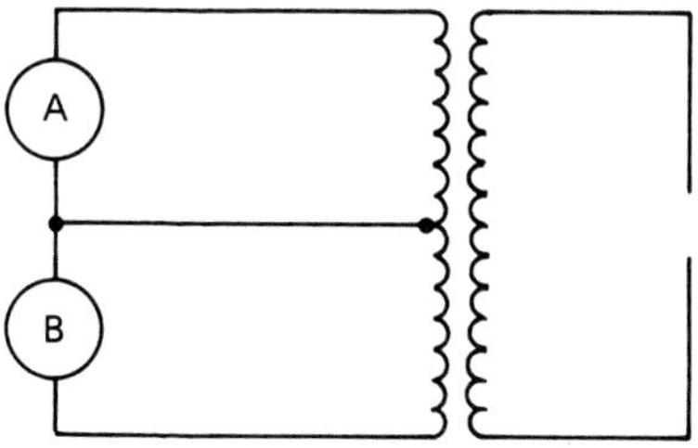


**Figure 5.2** Mono cannot be produced by paralleling A and B (a). Mono can be produced by connecting output transformers in series (not recommended) (b). Virtual earth mixing (c) produces a mono signal.

It is possible to connect transformer-balanced outputs in series but this is not recommended in anything other than an emergency.

You cannot combine MS and AB sources directly on the same mixer unless one of the formats is converted before they meet on the busbar. You do this by producing a minus S or minus B and panning the signals as shown in Figure 5.3.

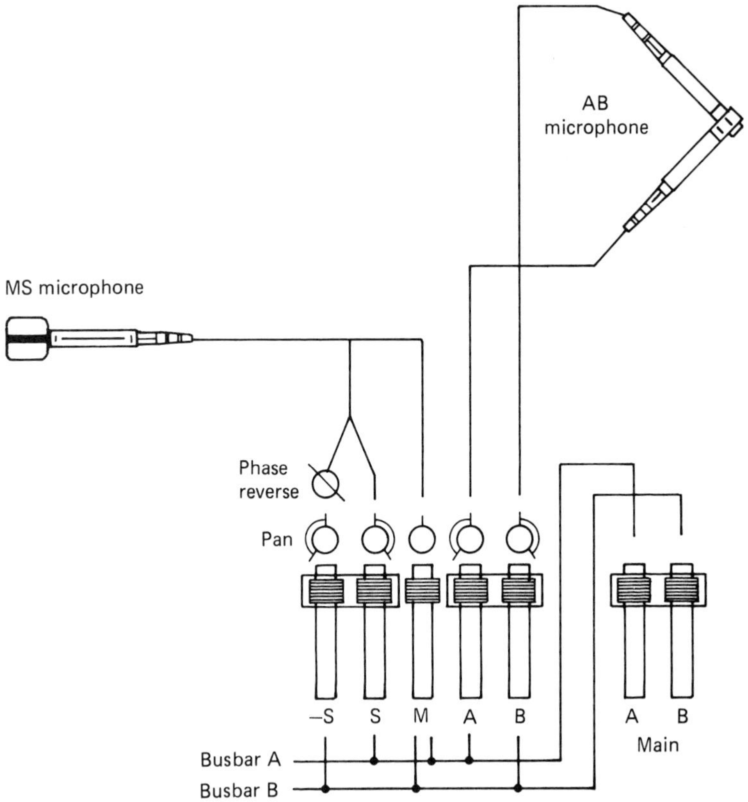


**Figure 5.3** AB and MS signals cannot be mixed directly. One set must be converted in the busbar mix.

It is vital that S and −S or B and −B track together. This is best achieved by deriving the minus component after the fader of the positive element, although fader ganging is an alternative.

It is therefore possible to retain width control of an MS microphone while mixing it with AB spot microphones, or, conversely, produce an MS output from combined MS/AB inputs.

## MS pairs

### *Tolerances*

Just as with AB, MS pairs of microphones need to be well matched, although sensitivity is not a major problem, provided they are corrected through the mixer. However, even small incompatibilities in the frequency responses will affect the stereo image, causing image spread in stereo, while leaving the central M sources unaffected. Because of the different 'sound' of microphones with different polar patterns, a definitive listening test is difficult. The more recent microphones are a lot more stable and will generally give a more accurate polar pattern than older designs.

### *Mounting MS pairs*

Having chosen your microphones, the next problem is mounting them. For true stereo imaging they must be coincident. Whereas in AB, microphone capsule spacing is mainly a worry for mono compatibility, in MS timing errors, even very small ones, cause cross-talk, which reduces the stereo separation. Some manufacturers offer piggy-back microphone mounts that give you coincident capsules, but only in the vertical plane. This means that the mount must always be in an upright position. You can rotate the S microphone independently of the M to pick up stereo atmosphere, although this would render them non-coincident. We will discuss this approach in more detail later. It is worth mentioning that you should put the figure-of-eight microphone above the cardioid to avoid the cardioid being shadowed.

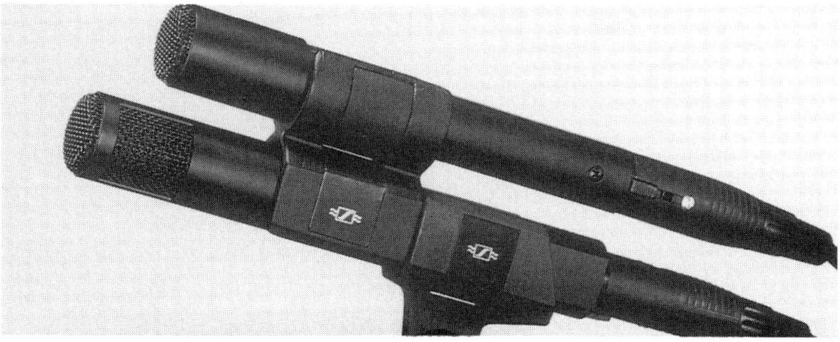

**Figure 5.4** Sennheiser MZD30 piggyback mount for MS pairs.

## *Choosing MS polar patterns*

The choice of MS polar patterns is just as important as with AB pairs. M as a cardioid and S as a figure-of-eight gives a linear stereo sound stage, but there are alternatives. A short shotgun such as the Sennheiser 415/6 as the M microphone, together with a figure-of-eight, is a powerful combination, but does have its limitations. The sound stage is certainly not linear, and there are areas of indeterminate phase that make judging the mono compatibility a little hit and miss.

Experiments have shown that the most balanced sound stage is produced when the S capsule's output is set 12 dB below that of the shotgun producing the M output.

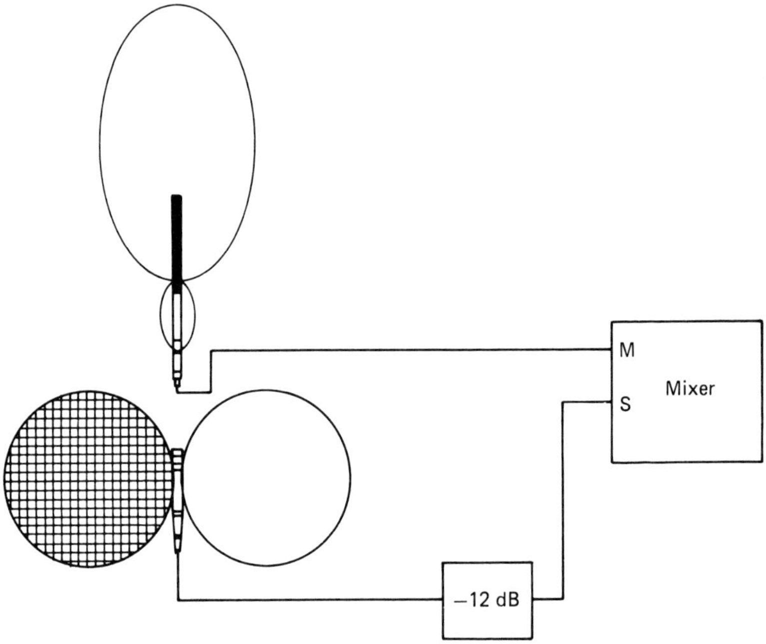


**Figure 5.5** Using a shotgun microphone in an MS pair. In practice the capsules should be as coincident as possible.

If both the M and the S are figure-of-eight microphones, they act as if they were crossed figure-of-eights in AB, although rotated by 45° with respect to the MS configuration. This is a very useful configuration for capturing ambient acoustics.

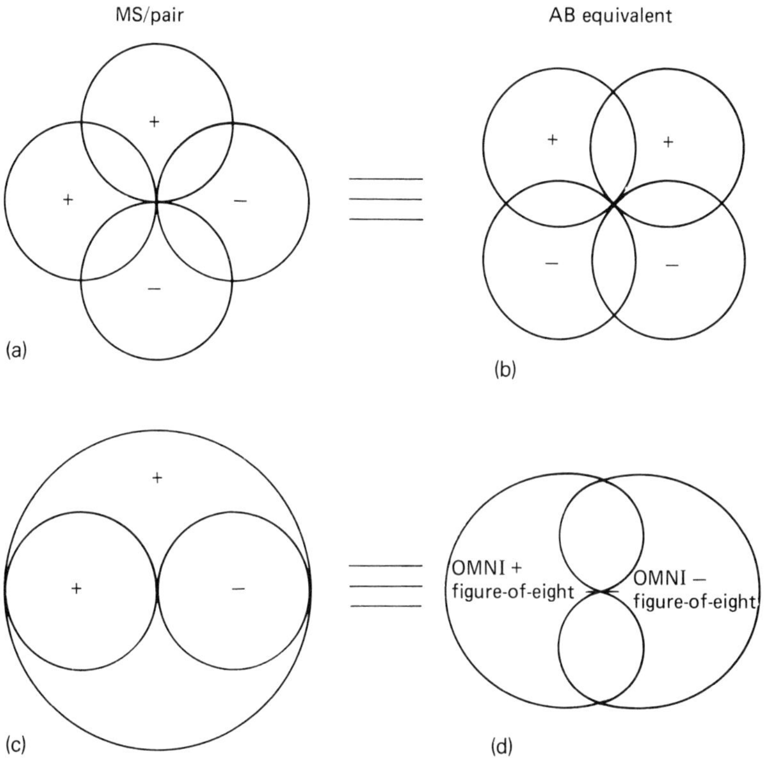


**Figure 5.6** (a) Omni with figure-of-eights at 90°; (b) crossed figure-of-eights; (c) omni with figure-of-eight; (d) back-to-back cardioids.

## *Side-fire stereo microphones in MS*

A number of the side-fire stereo microphones will facilitate both AB and MS working. As most come with variable polar pattern selectors you can choose the optimum microphones for mono compatibility. Because of their physical layout, they are coincident in the vertical plane and, with the accurate matching of the capsules, they are well suited to the tighter tolerances needed for MS working. One can normally expect excellent stereo imaging and accurate mono compatibility.

It should be very simple to set up a side-fire microphone as an MS pair, but somehow most of the manufacturers have got themselves into a mess with the markings on the microphones.

First you need to establish which capsule is the forward-facing member of the pair. If you choose the opposite one from the manufacturer, normally the lower capsule, the MS markings might as well be in Cyrillic! Switching one capsule between figure-of-eight and cardioid should solve any arguments as to which capsule is which, and you should be able to find some form of mark or LED showing the front face.

With the M (normally the lower) element set, move on to the S. It is critical that it be at precisely 90° to the M. Using the front marker as your guide, adjust the S capsule. Once the capsule angles are set you need to select the polar patterns you require. Most manufacturers supply additional boxes that not only have the pattern selectors, but also supply power to the microphone, via a special multiway cable. This means they cannot be used without the selectors.

Thanks to the miniaturization of the components, a number of manufacturers are now offering polar pattern selectors integral to the microphone. This makes the whole thing a lot more compact for the location recordist.

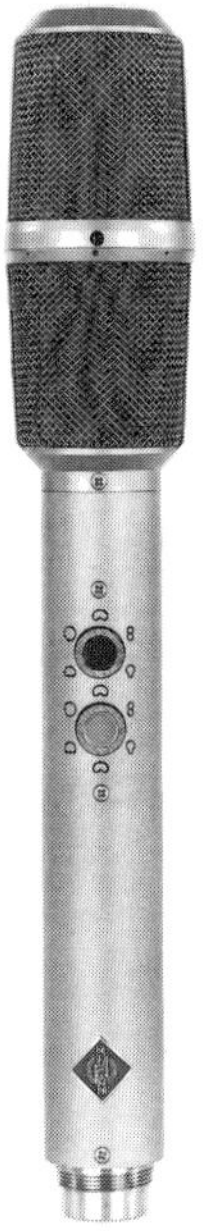

**Figure 5.7** Stereo microphone with integral polar selectors.

There are benefits to single stereo microphones that can be used as either an AB or an MS pair. If you decide that you do not require width control at this point, then using them as AB allows direct mixing with other AB sources and full stereo monitoring. If width control is deemed necessary then MS is configured without the need for a matrix.

It is worth pointing out that one of the real appeals of working microphones as MS is the ability to point them at the action. Unfortunately side-fire microphones preclude this. This limits the benefits obtainable with these particular devices.

### *End-fire MS microphones*

Television was initially attracted to MS miking because of the difficulty in aiming an AB pair. This led to microphone manufacturers offering single piece microphones, housing both an M- and an S-type capsule (see Figure 2.4, page 29).

Unlike the previously mentioned AB stereo microphones, these are end-fire. This means that boom and fishpole operators have a microphone that they can point at the action. It is necessary to align the S capsule correctly, i.e. with the positive capsule facing camera left, otherwise you will get a left-right reverse.

There are some limitations worth noting with this type of microphone. The figure-of-eight capsule is normally mounted behind the M, therefore rendering them near-coincident. Also the all-in-one approach means that your choice of M microphone is decided by the manufacturer. The benefit of selecting microphones precisely for each situation is a facility that is sometimes overlooked.

An inherent problem with any figure-of-eight type microphone is its natural susceptibility to wind noise. Any pressure-gradient microphone will show this to some degree. Bass tip-up can also be expected.

Most of the available end-fire microphones have the standard configuration of a forward-facing cardioid microphone coupled with a sideways-facing figure-of-eight microphone. A number come with their own matrix boxes, to convert the output to AB if required.

As with all equipment, some are better than others, but all are at the top end of professional sound equipment. Probably the best known MS microphone is the Neumann RSM190i or the later RSM191. This uses a short shotgun as the M, which undoubtedly gives more reach than its cardioid cousins. As you would expect, this combination does not give a linear sound field in stereo. The phase anomalies produced do make it sound rather wider than the acceptance angles on the matrix box would lead you to believe. It is nevertheless a microphone of the highest quality.

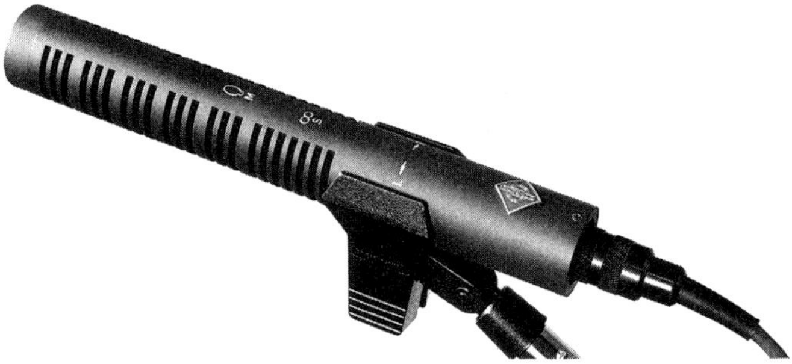

**Figure 5.8** Neumann end-fire stereo microphone.

### *Using an end-fire MS microphone*

To the uninitiated the end-fire stereo microphone mounted on a pole looks identical to its mono parent. However there are major differences. Its degree of movement is less and a number of the standard mono practices are denied you. The arc type of track is now a thing of the past, as the lobes of the figure-of-eight microphone should remain in the same horizontal plane at all times. But with some thought and a little practice, quite complicated set-ups are possible.

You use an MS microphone in exactly the same way as its AB counterpart and, as mentioned in Chapter 3 where we discussed moving AB pairs, microphone movements are more easily understood if you think in MS. Now, with the additional benefit of the end-fire capability, your operational limits become even more obvious.

The M signal is achieved by simply pointing the microphone directly at the action. In practice there are only a limited number of positions that give the best compromise between the camera shot, the lighting, and the sound. As stated before, the most important factors, after the source you want to record, are the sources you don't.

Map into your mind where things are and what sources you may need to avoid. Remember that you have a figure-of-eight microphone that may well end up facing directly into the crew working area. Try as much as possible to think of the way the shots will fit together. If you establish an off-camera sound, e.g. an open window, as being camera left on the wide shot, then it should remain on the left when you are recording the singles. It is probably best therefore to choose a sound stage that gives you a symmetrical sound field, however unreal that may be in relation to the camera angle. It is extremely disconcerting to have a stereo sound stage that is heavily biased to one side.

For example, how would you approach the recording of a piece to camera beside a busy road? The camera may well be showing you a scene with the road to one side of the frame. Reality would dictate that you match this and have the cars passing camera left. Although it is not wrong to record it this way, why not turn the microphone so as to balance out the sound stage? Cars would tend to enter camera right and exit camera left, enhancing the stereo effect and stabilizing the central dialogue.

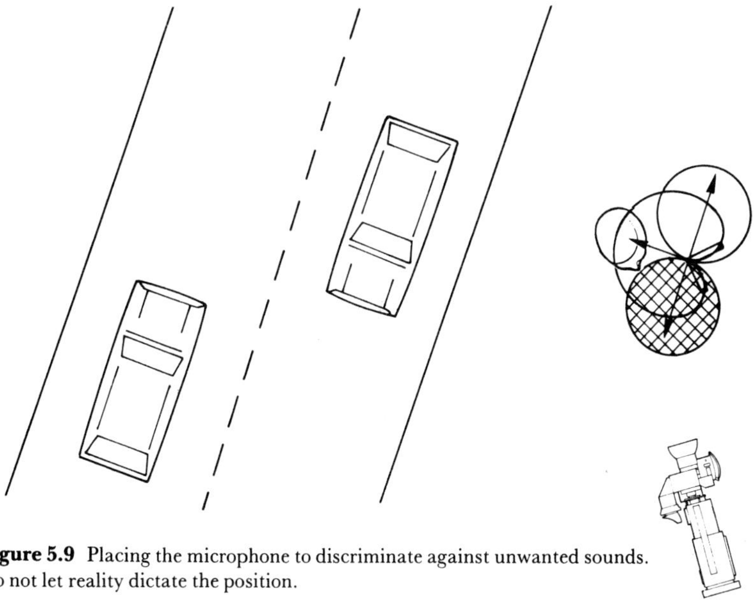

**Figure 5.9** Placing the microphone to discriminate against unwanted sounds. Do not let reality dictate the position.

### *Windshields*

Because of the pressure-gradient S capsule of MS microphones, they are inherently susceptible to handling and wind noise. Therefore suspension mounts and some form of windshield are recommended in almost every situation.

All manufacturers will supply specialist equipment, as many stereo microphones are slightly larger than normal mono microphones, but it is possible to obtain mounts that will allow you to fit them into standard windshields. Obviously it is critical that the microphone's body is correctly aligned, with the positive capsule facing camera left.

On location it is going to be necessary to re-align the microphone on a regular basis. Unfortunately the windshields are not designed with this in mind. Adjustment quickly becomes an annoying and time-consuming task; it is advisable to remove the microphone totally on each occasion, as turning it while fitted can damage the mount.

Some matrix boxes supplied with stereo microphones offer a −AB and −MS position that reverses the sound stage (see Figure 5.8). This should cut down the number of times you need to dismantle the assembly. Also if you are working with an MS input on your mixer, then phase-reversing the S signal will have the same effect. As many location mixers do not provide an input phase-reversing ability, keeping a phase-reversing lead in your kit will save you having to dismantle the windshield.

### *Split MS*

The value of this technique is a source of deep disagreement among operators. Because it in no way offers an accurate representation of a sound stage, many consider it unworthy of the name stereo, and in the true sense of the word, they are right. However, television is rarely about reality, only the illusion of reality. This technique gives you the power to offer a degree of width in situations where true stereo would be difficult, or impossible, to record.

Splitting the M from the S means that the phase relationship of the pair is totally arbitrary. Any image that results will be just an effect. What you gain is total control of the mono signal.

Nothing you do with the S microphone or the S gain will affect the mono signal in any way. The stereo may be false, but provided you keep the main sound sources off-mic to the S microphone, you should not get any major left-right anomalies between the sound stage and the pictures.

Consider the situation where two people are sitting, evenly spaced, around a circular desk, having an open discussion. A stereo microphone of either MS or AB could not cover this unscripted action without swinging the sound stage. Two stereo microphones might fare better but would still present the operators with problems.

One could cover the dialogue in mono and use a stereo pair as the effects microphones to record a wildtrack of the room-tone. This is a perfectly adequate way of adding stereo ambience but, as we discussed earlier, it has also opened out the mono. The quick and easy way to give the operator total control is to take the mono component away from the stereo pair.

When the MS pair is coincident, just as with AB pairs, you are recording level and not time differences. Once they are split you receive both. As M remains separate from S, no comb-filtering takes place. But as M and S are combined in the matrix to form A and B, the timing errors become significant to the stereo listener, as the time-of-arrival cues are now providing the dominant directional information. This means that, although in this example we use a figure-of-eight microphone for the S, there is absolutely no reason why any directional microphone could not be used. As it is critical that direct pick-up of the sources is minimized, a highly directional microphone may well be an advantage.

Whenever you are working MS you need to keep in mind the mathematical relationship between MS and AB. If, for example, the positive lobe of the S microphone was too close to the M so that for the most part M equalled S in both level and phase, then fading up S would not add ambience but merely pull the central image to the left (because if $M = S$ then $M + S = 2A$).

In practice, what you need to do is to configure the mixer to work with an MS input. The microphones covering the dialogue are routed to the channels designated as M. A single figure-of-eight microphone, or the S output of a stereo microphone, is routed to the S input. The placing of the S microphone is best discovered by listening tests. It is advisable to start with it at 90° to the action. The mono microphones cover the dialogue, just as in mono.

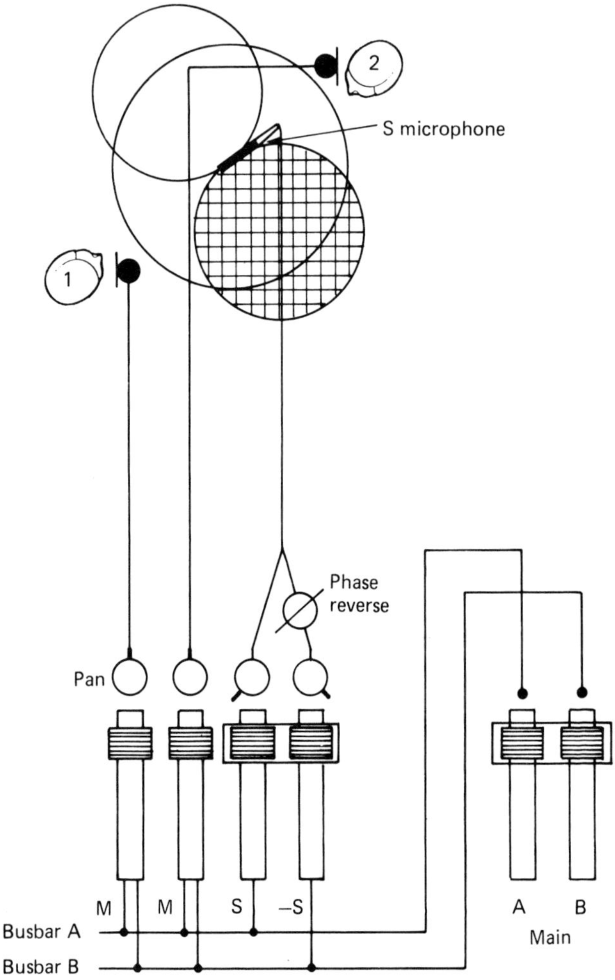


**Figure 5.10** Position the S microphone to minimize direct pick up.

The mixer then mixes the M and S signals to give an ambience. There should be no major left-right bias to the sound sources if you have correctly placed the S microphone.

It is possible to take this split procedure further. The M component can be made up of a number of mono microphones mixed, and even panned, exactly as in mono. For example, if the current trend of using hand-held cameras with participants on radio microphones is a problem for the recordist in mono, then in stereo it is a nightmare. Any attempt to pan the individual radio microphones to match the pictures rapidly falls apart as the camera moves across the action. Using a stereo microphone or pair for directional information just opens out the mono, probably negating most of the reason for using radio microphones.

An alternative would be to mix the radio microphones as a straight mono mix and have another microphone off the action as the S side of the mix. Again you would only be adding ambience rather than left-right information, but positional distractions are the last thing you would want in this situation.

As we discussed earlier, what we are after is the time-of-arrival differences between the M and S microphones. So the S microphone really ends up becoming little more than a delay device. This means that very much the same effect can be obtained without the need for an S microphone at all! Instead a delay of around 1–3 msec can be introduced to a split of the M signal which is then routed to the S input of a matrix.

Split MS gives the operator greater freedom on location but does limit the dubbing mixer to a mono-plus-ambience mix. If presented with true stereo it is always possible to MS mix out any unwanted positional discrepancies, while retaining the movement when required. You have to decide which is best suited to each situation.

## AB or MS?

The broadcast process is divided into seven main sections:

1. Miking techniques.
2. Mixing.
3. Monitoring.
4. Recording.
5. Editing.
6. Post-production.
7. Transmission.

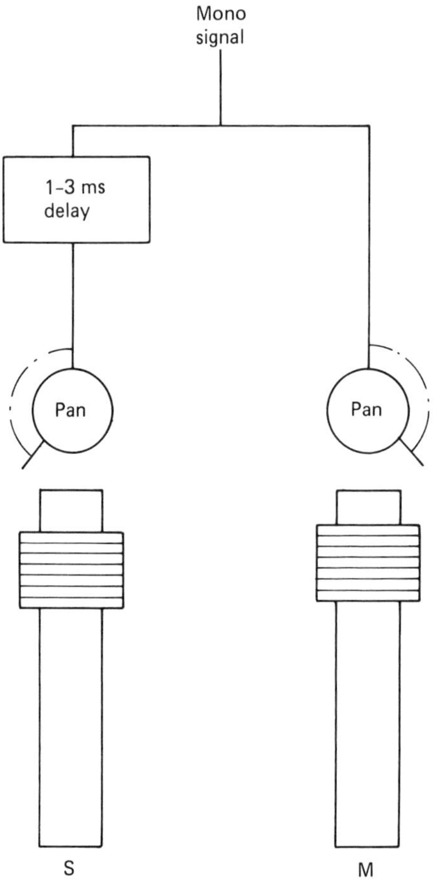


**Figure 5.11** Using artificial delay to produce pseudo stereo.

Any combination of six of these could be carried out as MS or AB. Loudspeaker monitoring of S, however, is essentially pointless. Meter monitoring of MS is extremely useful, but only in conjunction with an AB meter. The important thing to realize is that just because you decide to use MS miking, this does not in any way preclude the rest of the process from being AB. More to the point, just because television dubbing may well work in MS, this does not mean you have to record MS on site. Use MS and AB where appropriate.

## *Matrixing*

Changing between AB and MS is relatively easy. This chapter started with the simple mathematics, A + B = M, A − B = S, M + S = 2A and M − S = 2B. This altering of states is known as matrixing. There are a number of ways of achieving this. Probably the best way to understand what is happening is to look at matrixing through a stereo mixer using pan pots or grouping.

From the mathematics, one can see that to get an AB output from an MS input requires two M signals, an S signal and a −S

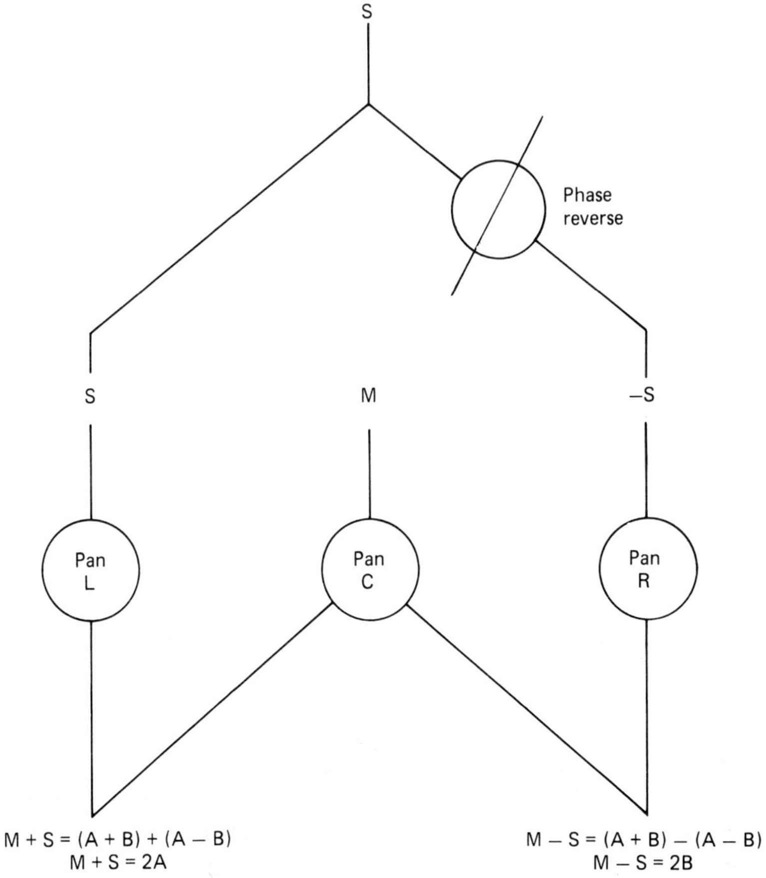


**Figure 5.12**

signal. Getting two M signals is easy. Take your M input and pan it to the centre so it feeds both the left and right outputs. Now turning to the S input, split it across two inputs and phase-reverse one of them. Looking at the in-phase S first, pan or switch it fully left. This now means that the left output has a mixture of M + S, i.e. 2A. For the phase-reversed S (−S), pan or switch this fully right. This gives you M–S on the right output, i.e. 2B.

The matrix we have now built will not only convert M and S to A and B but also the reverse. If one were to replace the M input with an A signal and the S and −S with B and −B, then M and S would appear on the left and right outputs respectively.

It is absolutely vital that the various inputs and outputs are equal in all respects, or correct matrixing will not take place. The tolerances are very tight, especially with MS inputs (see Chapter 1). If S and −S are not balanced it will introduce an offset. This will be most noticeable on mono sources, which will move as the S is adjusted. If B and −B are used, then any level differences will alter the width.

The accurate setting of S and −S (or B and −B) can easily be accomplished by panning both inputs to the same output and adjusting the relative gains to produce the minimum signal. Do remember to fade out the M microphone during the line-up and to re-set the pan controls afterwards.

To avoid having to control three faders for one source, it is desirable to source the phase-reversed inputs (−S or −B) from a point after the in-phase fader but before the busbar. Unfortunately no current portable mixer offers this facility. You would need to upgrade to a transportable. It is practical to use the fader ganging facility available on certain mixers to bring the control of B and −B or S and −S on to a single fader (Figure 5.13).

The major drawback of using a mixer to matrix is that it is clumsy and uses up channels that are normally in short supply on location. Stereo itself eats up faders, without the added complication of matrices.

Also, in many situations no level adjustment is required; in fact it could well be a positive disadvantage.

What is needed is a simple box that has two inputs and two outputs. If you put in MS, out comes AB and *vice versa*. A small number of manufacturers do now offer this type of equipment (Figure 5.14). However, as always, check out each, as strict

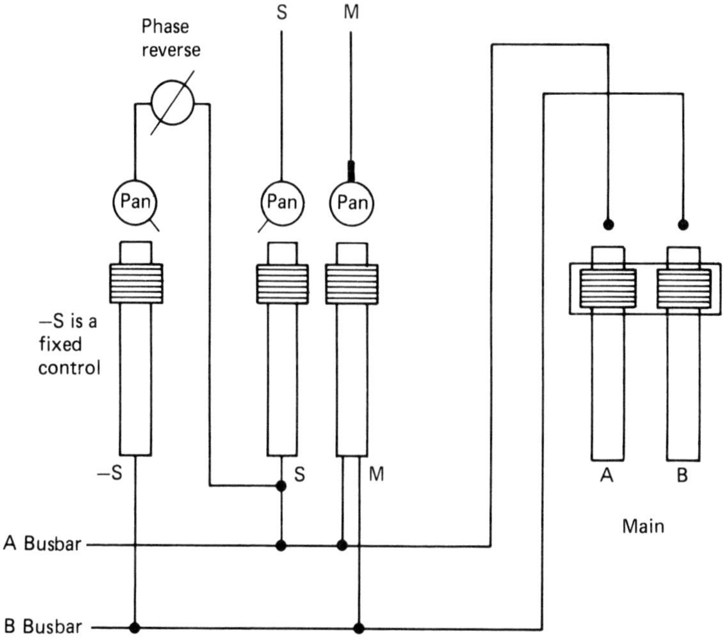


**Figure 5.13** The gain of –S must perfectly match S, then only the S fader need be used for width control.

unity gain throughout is critical. The less expensive types are based around transformers and are well suited to microphone levels, while the more expensive are active amplifier units that perform well with higher signal levels. If you intend to use microphones that require phantom power, be sure to specify this as many of the units do not pass phantom volts.

Many stereo microphones are supplied with their own matrix boxes. Normally these also house a battery to power the capsules. Some even offer an integral headphone amplifier. All the matrices have the facility to give either a straight MS output from the microphones or a matrixed AB. The more sophisticated offer a left/right reverse allowing 180° turns of the microphone, without having to resort to removing the windshield.

Audio Developments have produced a very versatile device that has two matrices, one in the programme path and one in the monitoring, all linked together with some clever switching and a stereo LED ladder meter.

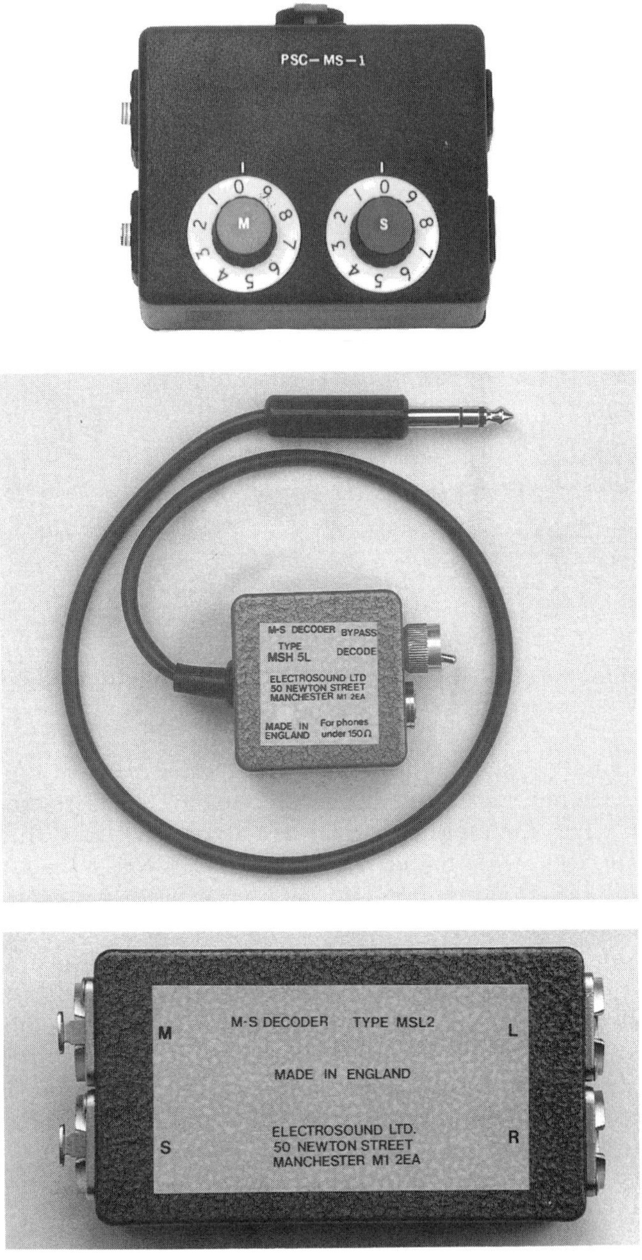


**Figure 5.14** Passive MS converters.

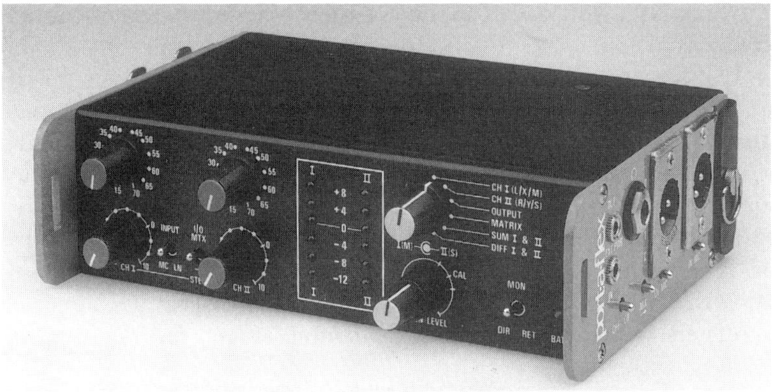


**Figure 5.15** Audio Developments ADO 066 stereo microphone amplifier.

Because of the higher tolerances required for MS working, it is advisable only to matrix those sources that really need it. For instance, matrixing every pair of effect microphones in a mix would be wasteful of equipment. Plan to MS the group as a whole, thereby separating the individual microphone balancing from the width manipulation.

## *MS mixing*

When using AB pairs, width manipulation is achieved by physically adjusting the angle between the capsules or by narrowing the image using the pan pots. However, the first is basically a set-up type adjustment, rather than a dynamic one, and will alter the mono response, while operating two pan pots takes both hands and is therefore not very satisfactory.

Because television needs to produce as complete a programme as possible on location, some form of dynamic width manipulation is necessary. MS mixing is designed for this purpose. Remember, even if your source material is AB, you can matrix it to MS for mixing.

This does not mean that MS mixing can rectify poor microphone techniques. The most important step is always that of placing and angling the microphones themselves. Once this is achieved you can move on to the mixing. Remember that the mono is derived from the physical setting of the capsules.

Nothing that you do with the S signal affects the mono mix in any way.

The first thing you notice is that standard audio fader law is not well suited to MS sources. Very little appears to happen until the M and S gains are nearly equal, then suddenly everything goes very wide with even small increases in S. This means that accurate fader tracking is vital. With inherently wide stereo sources it helps if, around the edit point, you lead with the M fader, trailing the S by 5 dB. As you reach the desired level with the M, the S is gently added until the correct ambience is achieved. When leaving a wide sound stage, reverse this by leading with the S.

With crossfades between MS sources, the fader law can actually work for you. Dropping the S by just 5–10 dB can effectively mono the source. Therefore as you approach the crossfade, gently drop the S from the first source and make a mainly M-to-M crossfade, with the second S lagging by the same 5–10 dB. Aim to cross all four faders at the same time, although

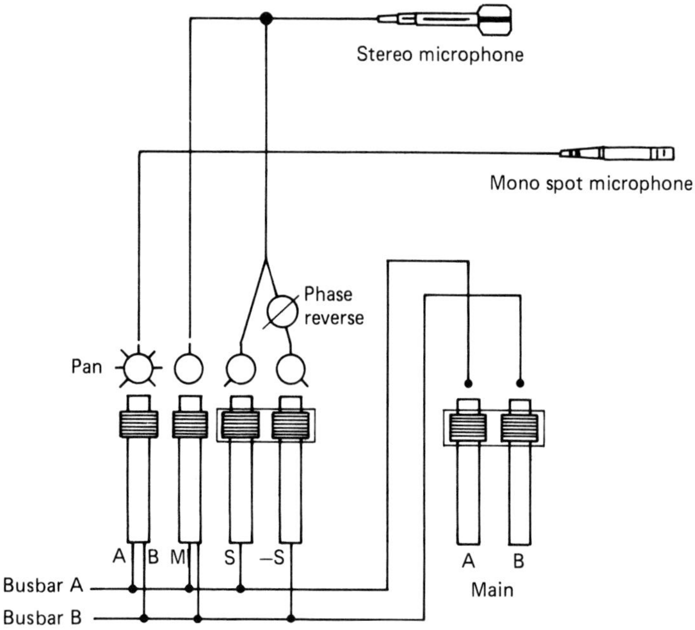


**Figure 5.16** If you need to pan a mono source then the busbar must be configured for AB.

keeping the 5–10 dB separation between M and S until the second source has fully taken over. These techniques are designed for linear faders but most portable mixers, even those configured for stereo, have rotary faders which, as you will discover, are not well suited to MS mixing.

There may be times when you need to mix a combination of MS and AB sources: perhaps panned spot microphones with a covering MS pair. This is perfectly acceptable and can be very effective. It is advisable, however, not to set up conflicts between the images. There are some technical problems that arise and these are dealt with in the section on matrixing. The important thing is that your busbar is working AB.

Finally, if you find yourself using either mechanically or electrically ganged MS faders, you should ask yourself why you are not working AB. There is little point using MS unless you require width control.

# 6

# Operational considerations

## Wind noise and rumble

We all remember that figure-of-eight microphones are prone to wind noise and rumble. Stereo theory has shown that all stereo microphones contain both forward-facing directional and sideways-facing (figure-of-eight) elements. It is this figure-of-eight element that will be prone to wind noise and rumble. The degree to which an individual microphone will be prone to these problems will of course depend on its own characteristics, but we can be sure that as we include more of this sideways signal in the microphone's output, the worse it will suffer.

Operators have reported that some fishpoles are prone to transmit excessive handling noise through to stereo microphones. Since the fishpoles concerned have worked satisfactorily with monophonic microphones, it seems that, again, it is the figure-of-eight element in a stereo microphone that causes the trouble. It would seem wise to check out existing fishpoles before committing them in the field to stereo.

### *Minimizing wind/handling noise*

One very good argument for working MS when recording in the field is the potential for putting bass cut into the S output of the microphone. Since it is this output that contains the figure-of-eight element, bass cut here will be most effective. There will be no effect on the monophonic output since this contains no side element. It can also be argued that the effect on the stereo signal is going to be minimal since extreme bass tends not to be locatable with any degree of precision.

## Reporters and stereo mics

It is probably wise to avoid giving a reporter a stereo microphone to hand-hold. Reporters have enough to think about without worrying whether the microphone has the correct left-right orientation, or is even pointing up and down. Reporters have also traditionally been trained to hold microphones so that they maintain the same spatial relationship with their mouth when turning – in other words the microphone turns if the reporter turns. This is exactly what we do not want in stereo. If this were to happen, the background acoustic would spin unaccountably, and, what is more, in exactly the worst possible direction. A little thought will show why this is so. Should the reporter turn to his left (camera right), the microphone will turn with him, bringing sounds originally to his left, and now within his looking room, behind and so off-mic. At the same time, objects off to his right, and now possibly out of shot, will be centre sound stage. Figure 6.1 illustrates the point.

## Phantom power supplies

All capacitor microphones take power, and some stereo microphones take a lot of power. Phantom powering involves resistors of a few thousand ohms in series with each leg of the signal circuit within the power supply, often as an integral feature of a mixer. Normally the voltage drop across these resistors is insufficient to cause any problems, but some stereo microphones draw enough current to drop the voltage available at the capsules to below that at which the microphone will function satisfactorily (Figure 6.2).

If this drop in voltage just caused a corresponding drop in the output level, we could perhaps live with it. Unfortunately, the symptoms can vary from loss of just one side of the microphone's output, through to total loss of coherent output combined with

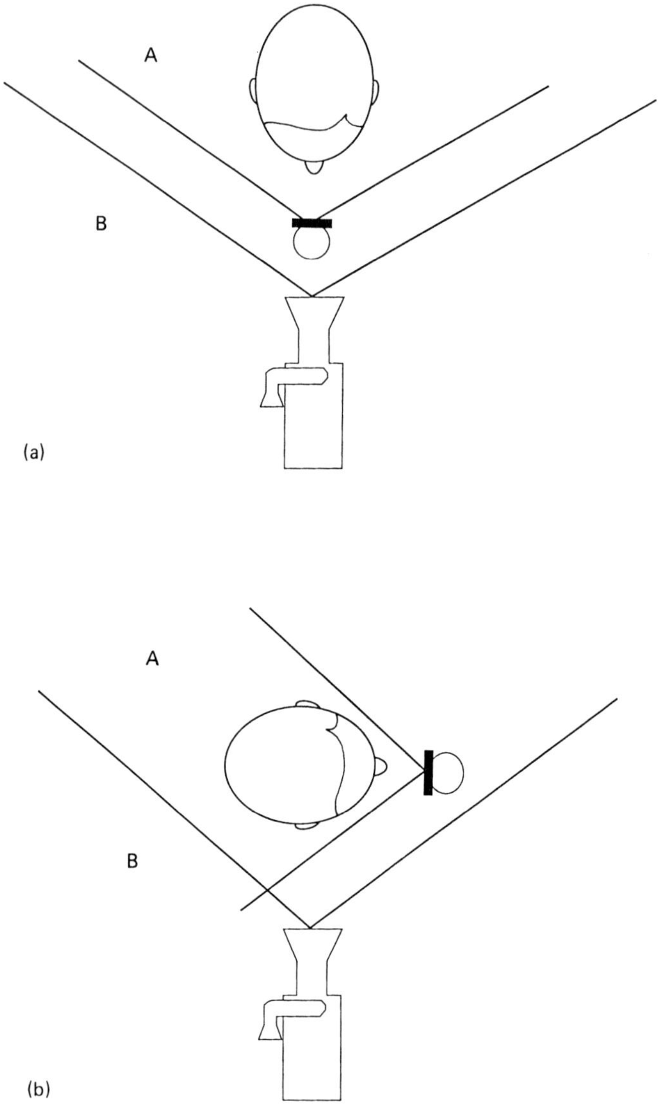


**Figure 6.1** Both camera and microphone points of view correspond (a). Secondary sound source (A) is both in vision and in its corresponding position in the sound stage. Secondary sound source (B) is both out of vision and off-mic. In (b) the reporter has turned while holding the microphone. Secondary sound source (A) is now on the opposite side of the sound stage to its visual image. Secondary sound source (B) is still out of vision but now well centred in the sound stage.

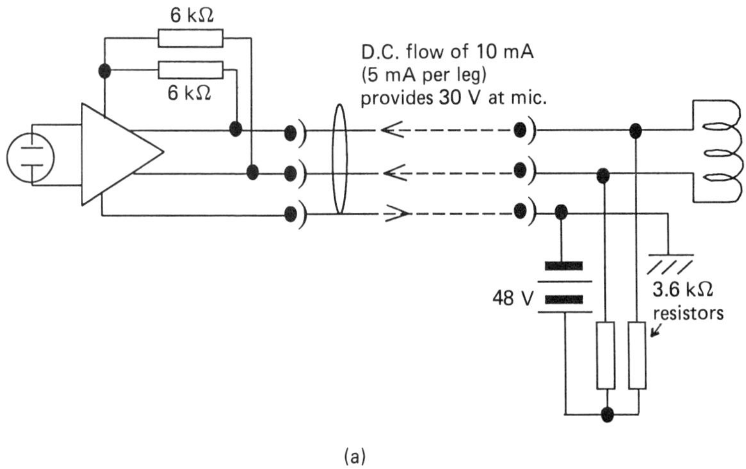


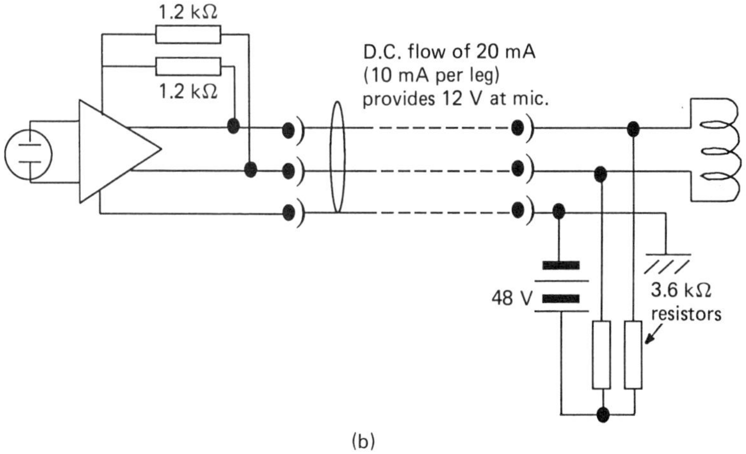


**Figure 6.2** Traditionally, phantom-powered microphones have drawn little current from their power supplies (a), leaving sufficient voltage at the head to power up the internal amplifier and polarize the capsule. Stereo microphones can take more current (b). Some traditional power sources may have too high an internal resistance to provide sufficient working voltage within the microphone head.

what sounds like pink noise. Some operators have reported that this loss of output on one side is dependent on which way round the microphone's two outputs are connected to the mixer, indicating that small differences in resistor tolerances within the phantom powering arrangements are the deciding factor. The only safe course of action is to try the microphone out with its mixer/phantom power supply before going off into the field.

Phantom powering usually requires 48 volts. Most field mixers operate on about 12 volts, so there are voltage multiplier circuits to make up the shortfall. The drain on the batteries when stereo microphones are in use can be considerably greater than with mono microphones. Expect to get through batteries quicker than when working in mono.

## Colour codes on leads

Traditionally in broadcasting, yellow leads have meant phase reverse, yellow PPM needles have indicated S signal levels, and red has indicated the left side of a stereo pair. BEWARE! One microphone commonly available, switchable between AB and MS working, has red and yellow output leads. They do not indicate left and difference respectively. The red lead is the right or side output, the yellow the left or main. Moral – check and then label the leads unambiguously.

## Matrix box terminologies

Matrix boxes often have gain controls for the S channel when working MS. There are differences between how manufacturers designate the settings of this control. Commonest forms of labelling are dB gain with respect to the M channel, acceptance angle and on-axis angle. The term dB gain is not of much practical use in the field, and the other two terms do not mean the same thing.

On-axis angle means the angle formed by the axes of the AB equivalent pair of microphones, whereas acceptance angle means the angle through which the sound source has to move in order to go from fully-left to fully-right. This angle is determined by the dead points in the polar response of each of the AB equivalent microphones (see Figure 3.12, page 44).

## Reversing the fishpoles

A simple point, but easily forgotten in the stress of recording in the field, is that the left/right directionality of a stereo microphone will reverse when you change from poling the scene from above to poling it from below (Figure 6.3). Some microphone matrixing boxes have a '−S' switch which will effectively reverse the left/right directionality. Failing this you can, if working with an MS pair, put a phase-reverse lead in the S output of the microphone before it is matrixed to AB. If working AB throughout, then either change the microphone inputs over, or swing the pan pots on the mixer over.

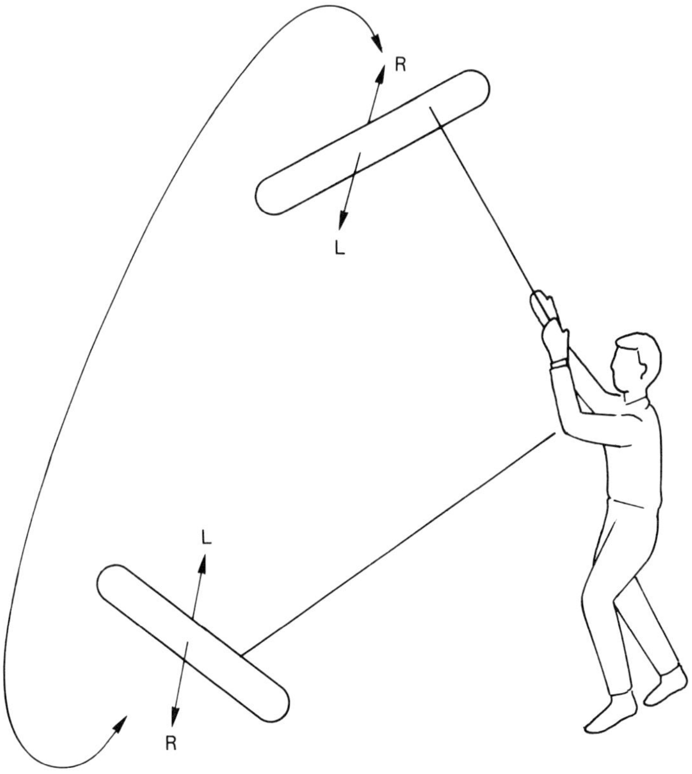


**Figure 6.3** Remember that left and right will swap over when changing from poling from above to poling from below.

## Compressors and limiters

We all like to have compressor/limiters as a backstop when recording in the field. Are we sure, though, that the two-channel limiters are ganged?

Unganged dynamic gain control devices can cause image shifts in the stereo picture. There are a number of reasons for this. Firstly, if the devices have different threshold levels, the image will shift away from the device which comes into action at a lower level. If the devices are ganged then both will act in concert.

If they have different attack times, the transients will be affected, with the image of transient-dependent sounds (plucked or struck instruments, for instance) moving towards the side of the image having the compressor/limiters with the longer attack time.

Release times are also important. If one half of the compressor/limiters pair has a longer release time than the other then the image will move away from that channel during the release time – a classic case of image wander!

## Wildtrack

Sound recordists have always been aware of the need for wildtrack, essential if sound post-production staff are to have a chance of saving their hair! In stereo (and even in mono if there is a chance of the company deciding to go stereo in post-production if sales hang on the decision), it is imperative that plenty of wildtrack is recorded, and from a variety of well documented positions. Atmosphere changes are noticeable enough in mono: in stereo they are a hundred times more so.

The dubbing mixer or picture editor must not only have the opportunity to 'sweeten' the sound, but may well also be faced with creating a saleable stereo programme from something that may have been shot in stereo, but with an initial view to marketing in mono only. Doubtless this ambiguity will vanish in time, but at present many programmes are made with the intention of dubbing in mono, but marketing constraints result in the post-production mixer being asked to produce a final result for sale in stereo.

Even if the programme is conceived in stereo, the dubbing mixer must have options open in post-production. Record a number of wildtracks from a variety of perspectives and angles to cover all the possible envisaged edits. Yes, this is a counsel of perfection, but when the only person likely to follow a production through from start to finish is the director, it is important to present your colleagues in post-production with the widest of options; after all, do you really want them resorting to the effects library rather than using your hard-won reality?

## Headphones versus loudspeakers

Any sound recordist who has compared the difference between listening on headphones and on loudspeakers will know there is no comparison. Headphones simply do not give the 'crossed' signals on which true stereo relies. True, modern 'closed' headphones exclude unwanted sounds, and experienced recordists have learned to distinguish between the off-tape delayed sound and direct sound leaking past the ear muffs. But headphones are not a substitute for a pair of loudspeakers in a friendly listening environment. Perhaps the worst shortcoming of headphones is the difficulty, in the field, of determining whether or not a stereo pair of microphones are in phase. To check this, switch the monitoring to mono with a central sound source, and phase-reverse one channel. If the sound becomes thin and reedy in quality then the channels were originally in phase.

There were occasions in mono when a pole operator, working to a sound supervisor, would not wear headphones, relying on their experience, and that of the supervisor. In stereo it is essential that the pole operator has full stereo monitoring, as the extra dimension of stereo imaging makes previously simple moves considerably more complicated.

As regards judging the stereo image width, this is virtually impossible with headphones. The only safe guide is learned experience.

Be aware that most field video recorders do not provide a stereo headphone output. It is usually mono only and switchable in just about every combination except stereo. There is a lot to be said for monitoring the output of the field mixer for most of the time, dipping into off-tape on occasion to check the recording. The other reason for monitoring the mono off-tape feed from the field recorder is that this will at least give you an indication of mono compatibility, and headphones cannot give a true representation of the stereo image.

# 7

# Mixing, metering and monitoring

## Control and monitoring

Working in stereo, especially MS, demands a far higher specification for all our equipment. You can use a large selection of mono equipment to produce stereo. But you must have a stereo mixer with the following facilities:

1. Pan pots.
2. Two meters.
3. Limiters with switchable ganging.
4. Stereo monitoring.
5. Derived mono monitoring for compatibility and phase checking.
6. Gangable main fader.

In addition it would be very useful to have:

1. Input ganging.
2. Input selector switches – left/right or panned.
3. Offset controls.
4. Monitor matrix.
5. Output matrix.
6. Stereo/mono tone generator.
7. Stereo auxiliary return.
8. Visual out-of-phase indication.
9. Twin movement meters.
10. Stereo faders.

If you are working in AB then fader ganging is very helpful, if not imperative. If you do not have this feature, then it may well be advisable to set your channel faders at an optimum setting, and ride the gain on the ganged main fader. Otherwise there are likely to be unpleasant image movements as you attempt to track the faders simultaneously.

### *Line-up of AB pairs*

Setting the relative gains of the stereo pair is extremely important. It can be achieved in the following manner. Set the faders to their normal positions. Make sure that any pads, bass cuts etc are identical on each channel and take the limiters out of circuit.

Adjust the microphones so both are pointing forwards and speak directly into them. Adjust the coarse gain controls so each meter peaks roughly the same. Use the faders themselves to fine-tune this. Once you are confident that a mono source will appear in its correct central position, set the acceptance angle of the microphones. As long as you track the faders together the image will retain its integrity, regardless of the gain of the system. The limitations of this system when using unganged rotary faders should be obvious.

A far better way would be to use a mixer like the SQN4S or the AD260 which have a fader-ganging facility. These need to be set up a little differently. As before, make sure that any pads, bass cuts etc are identical on each channel and the limiters are out of circuit. Now switch in the ganging circuit. In this mode, channel 1 fader controls both inputs 1 and 2. The pan pot on channel 1 acts as an offset control which allows very accurate calibration between the inputs. Again get someone to speak directly to the microphones and adjust the offset control so that the meters peak the same.

Note: It is necessary for the pan pot on channel 1 to be in the pan position. Channel 2's fader and pan pot are inoperative.

### *Using pan pots to control width*

The width of an AB stereo pair can be adjusted using the channel pan pots. As you move the pan pots towards the centre

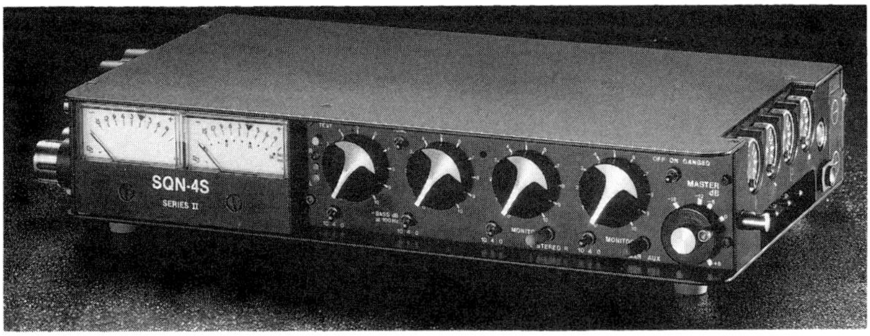


**Figure 7.1** Portable stereo mixer with fader ganging facility.

you are decreasing the amount of difference, S, between the channels, therefore it is tending towards mono. This method will not give you the ability to increase the effective width beyond that of the microphones themselves, nor is it a very dynamic control. This means of width control should be used only as a last resort. It is far better to fully pan, or preferably switch, the microphones to their appropriate groups. Set the width by correct microphone placing, or actively adjusting it using an MS matrix.

As the SQN4S uses its pan pot as an offset control in the ganged mode on channel 1 you are unable to electrically adjust the stereo width in this mode.

### *Line-up of MS pairs*

The setting of an MS input is more problematical. Level differences affect the perceived width rather than the offset, so there are no meters to balance. The only thing to do is to set the M level on input 1, then reverse the inputs and line up the M to the same level on input 2. Reset and then gang the faders if required. If your mixer has a phase-reverse selector, it is useful to remember that phase-reversing the S input gives you a left/right reverse, which can be a very useful feature.

### *Manufacturers' width controls*

Although not yet available on smaller desks, some manufacturers offer separate width controls on stereo sources. Being purpose built they have a linear response and usually cover the range from mono, through stereo, to so-called wide stereo. They work, but operationally width needs to be a dynamic, rather than a fixed control, and as you need one hand on the fader and the other on the width control, full stereo mixing is difficult.

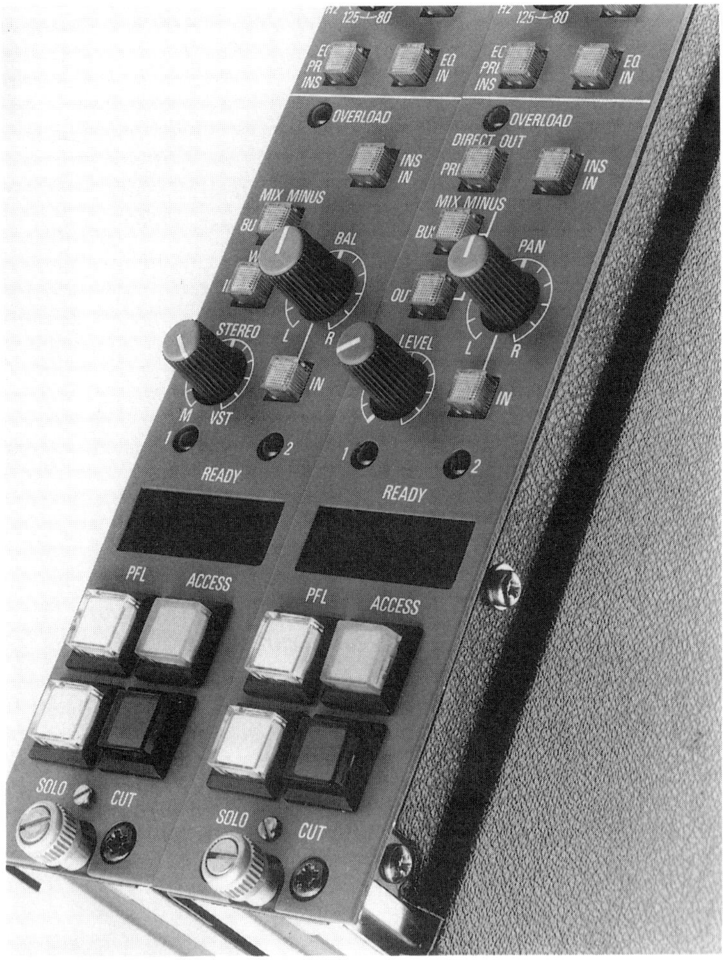


**Figure 7.2** Manufacturers' width control.

## ***Panoramic potentiometers***

If we had a simple system that kept the voltage constant across the sound stage, then a mono source panned from left to right would not change in level for the mono listener as the image moved. The stereo listener however would notice a drop in level for central images, as the combined power of the system would fall by half in the centre. If we keep the power constant across the pan, the stereo listener would be happy but the mono balance would be 3 dB high on central images.

Most designers opt for a mixture of the two systems. It is a compromise, and operators need to be aware of its effect on the two mixes.

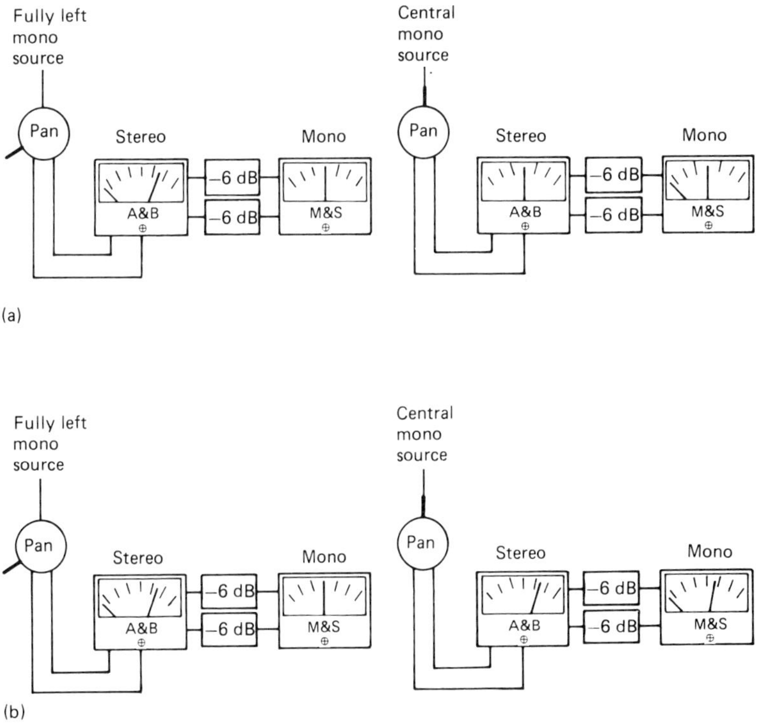


**Figure 7.3** (a) Constant voltage pan pots. Stereo level drops while mono level remains constant. Note 6 dB pad in derived M and S (ITV standard). (b) Constant power pan pots. Stereo level remains constant while mono level rises. Note 6 dB pad in derived M and S (ITV standard).

### *Stereo levels*

If you add together two mono sources, of the same level, the resulting signal will be 6 dB higher. This is because the signals are in-phase or coherent. If you take two signals with the same level, but with a totally random phase relationship, or non-coherent, such as white noise, the resulting signal will be only 3 dB higher. Unfortunately, stereo signals within a programme will sit somewhere between these figures.

In practice it is necessary to attenuate the derived mono by a fixed amount so as not to over-drive equipment in the mono signal chain. Any figure you choose can only be a compromise, as the actual level increase will be dependent on the stereo content of the programme.

### *BBC line-up*

It was discovered by BBC listening tests that if you put a 3 dB pad in the derived mono signal, i.e. M = (A + B) minus 3 dB,

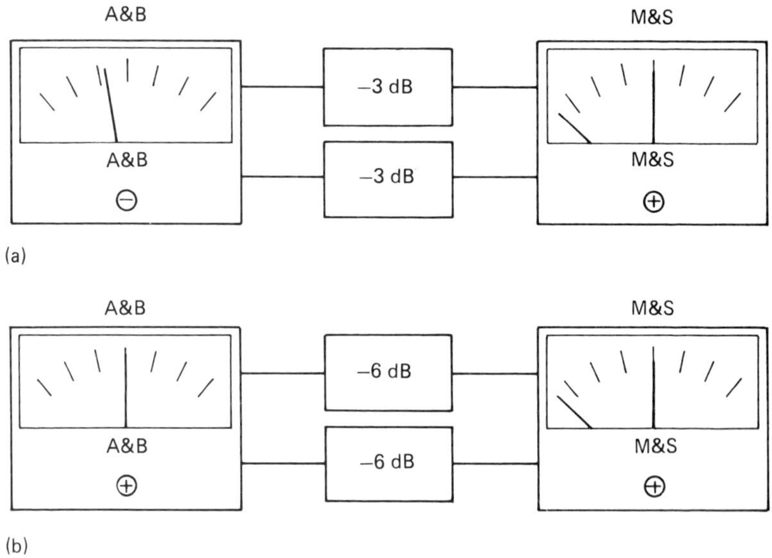


**Figure 7.4** (a) BBC specification. M = (A + B) −3 dB. (b) ITV specification. M = (A + B) − 6 dB.

the relative sound levels 'sound' similar in both stereo and mono on most types of programme. Also AB and MS meters give a good match. On line-up M then reads 3 dB higher than the individual A and B channels. Note that the same attenuation must be applied to both the M and S meters.

On location you may rarely have a mono meter. But the BBC insists that the mono signal, A + B minus 3 dB, should not exceed PPM 6, so the A and B meters should peak around PPM 5 or 3 dB below peak. However, if the source is extremely one-sided, A or B peaks of PPM 6 are acceptable.

### *IBA/ITV line-up*

The IBA opted for a different standard. AB and MS meters will peak zero (PPM 4) on tone, i.e. M = A + B minus 6 dB. However, the mono loudspeakers will probably be to BBC specification, (A + B) minus 3 dB, to give an accurate acoustic balance, stereo to mono, while the summed mono derived at the transmitter will be (A + B) minus 4.5 dB.

ITV's 6 dB pad would lead you to peak to PPM 6 on A and B for a centre image, but any programme that was extreme left or right would then be recorded 6 dB below peak in mono.

### *Stereo tones*

So as to identify tapes as being stereo or mono, a new standard of test tones has been introduced. Unfortunately BBC and ITV use different standards.

At first, different frequency tones were proposed, but this was quickly rejected as it could lead to incorrect line-up if the recording machine did not have a flat frequency response. Also phase checks between different tones are meaningless. The IBA finally settled on a system of 1 kHz tone on both channels but with a break of 0.25 seconds on the left channel every three seconds.

The BBC took this further, with a system having one break on the left channel and two on the right. This allows for a truly astonishing amount of information to be extracted.

| **You hear**: | **Indicates:** |
| --- | --- |
| Tone with one break | Left channel of a stereo tape |
| Tone with two breaks | Right channel of a stereo tape |
| Tone with three dips | Mono from a stereo tape |
| Tone with three peaks | Phase reversed |
| Tone that breaks left and right | Stereo |

### *Metering*

The main concern of every recordist working in stereo should be mono compatibility. Yet, if you are working on headphones, and have inadequate meters, major mistakes are likely to occur. At present we tolerate an abysmally low standard of metering on location equipment. Stereo operations will demand considerable improvement. In all stereo work the important factor is the relative levels and phase of the two signals.

The majority of currently available portable mixers only have separate A and B meters. They are useful for tone line-up but on stereo signals it is difficult to gain any useful information, other than over-modulation. The level of the derived mono signal is left to guesswork. But most importantly, these meters give you no indication as to the relative phase of the signals.

If MS signals are processed then you can judge the phase from the meters but no left/right information is presented.

Equipment that offers meters to examine relative errors is scarce. The Stereo Nagra's Modulometer ballistics may be unusual but at least it is a dual meter. A few mixers have bargraph-type meters that will show interchannel differences, but none has twin-movement dual meters.

### *Audio vectorscopes*

If possible, one of the new breed of audio vectorscopes would be a welcome addition on larger location set-ups. The units are transportable, rather than portable, but give you comprehensive metering.

They resemble an oscilloscope and display a trace that gives a pictorial representation of the phase relationship between the A and B signals. Despite looking rather muddled at first glance, it

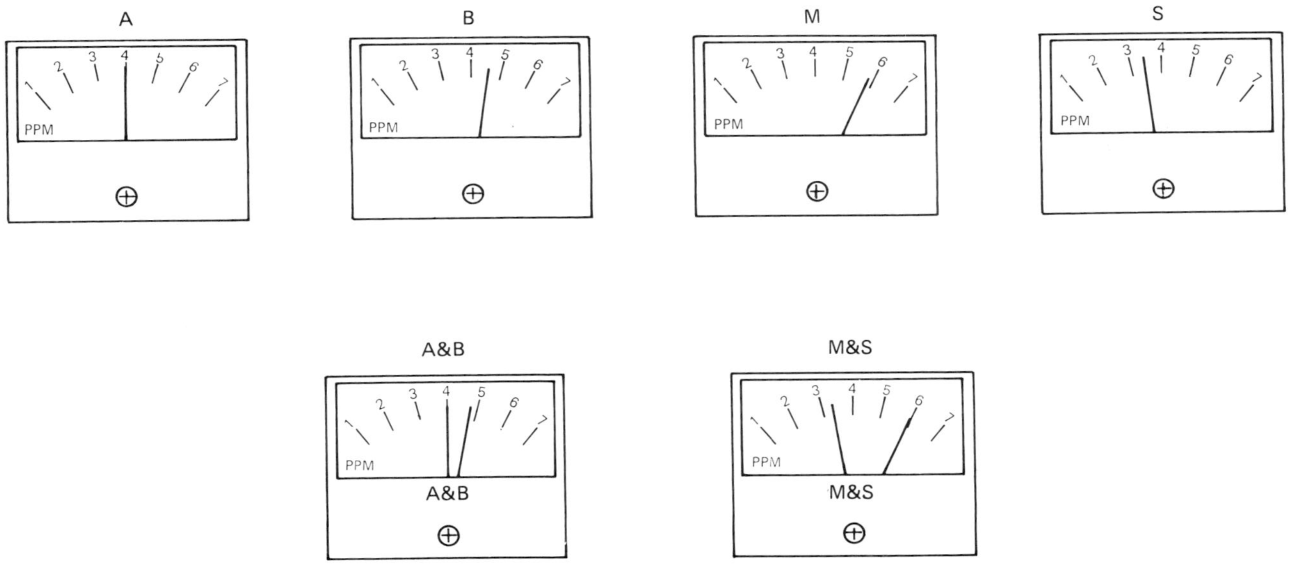


**Figure 7.5** Separate meters are confusing and give little interchannel information (a). Dual movement stereo meters give level, offset and phase information (b).

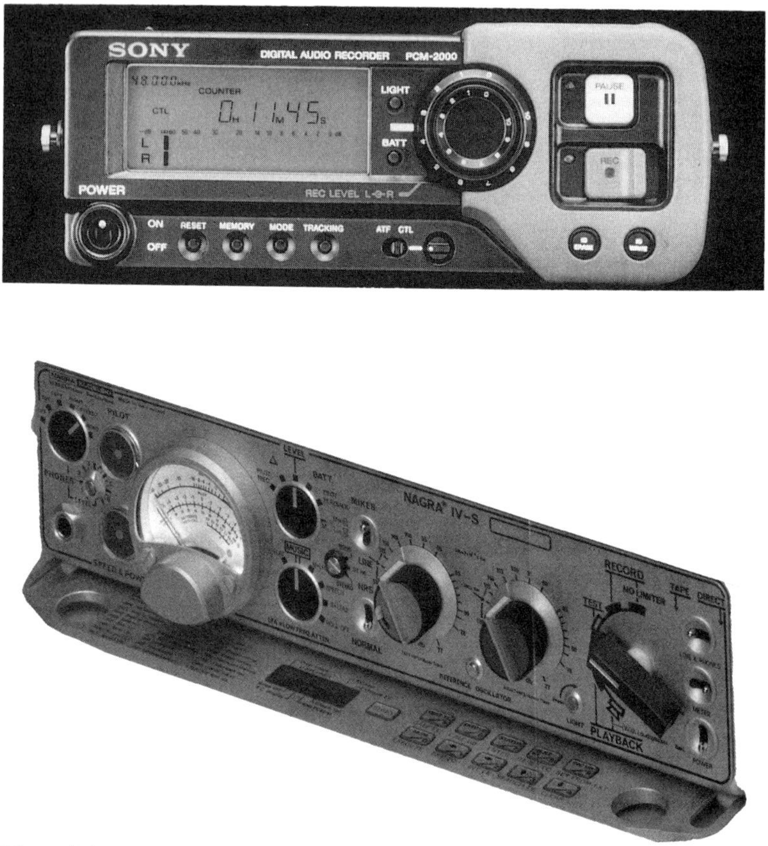


**Figure 7.6** Dual meters are essential for monitoring stereo levels.

is well worth persevering, as within a few minutes the value of this form of monitoring will become apparent. These are not quantitative meters, they give no absolute level information, but many have additional bargraph meters switchable between AB and MS.

### *In-picture meters*

If one is working in an outside broadcast vehicle, then an 'in-picture' meter, showing AB and MS plus a phase meter, would be worth experimenting with.

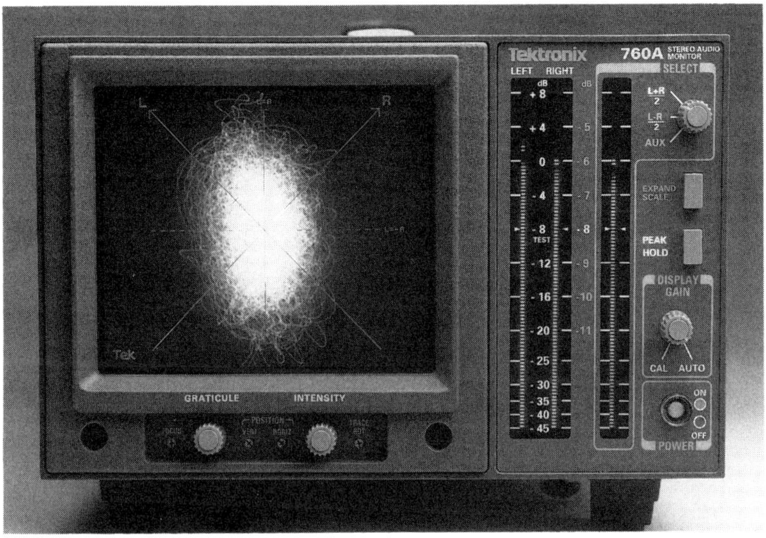


**Figure 7.7** Audio vectorscope.

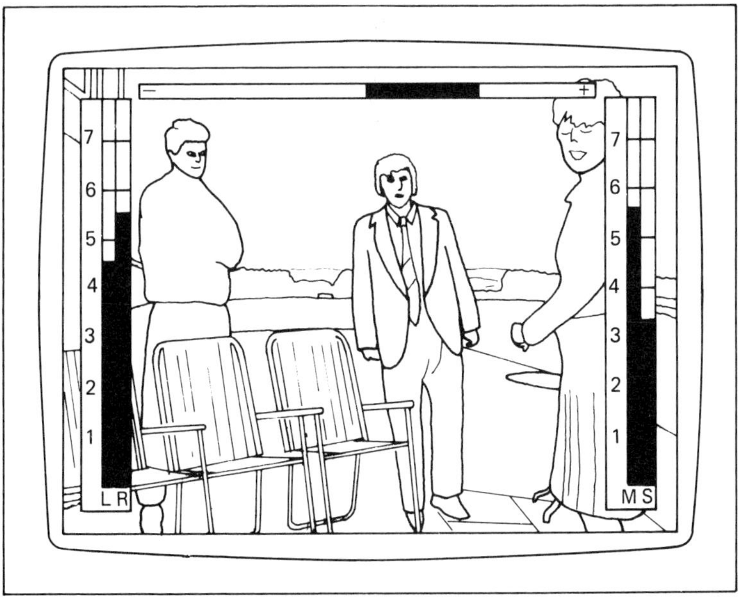


**Figure 7.8** In-picture stereo meter.

# 8

# Recording sound on video recorders

## The recording process

Although the basic principles of magnetic recording hold for both audio and video signals, there are significant differences in how the principles are employed, and it is as well to understand these.

These basic principles are reasonably well known. A plastic tape, coated with finely divided magnetic powder, is passed at a linear speed in front of, and in intimate contact with, a pair of magnetic poles in the form of a recording head (Figure 8.1). Audio currents flowing in the head coils cause corresponding fluctuations in magnetic flux, and the particles of magnetic powder are accordingly magnetized to varying degrees, depending on the strength of the audio signal current flowing in the head coils as the particles in turn, having passed in front of the pair of poles in the recording head, leave the influence of the magnetic field.

### *Traditional requirement for bias*

Unfortunately, this process of translating the audio signal into a magnetic imprint on tape is non-linear, and to overcome this a high frequency bias signal has traditionally been added to the audio signal in the record head (Figure 8.2).

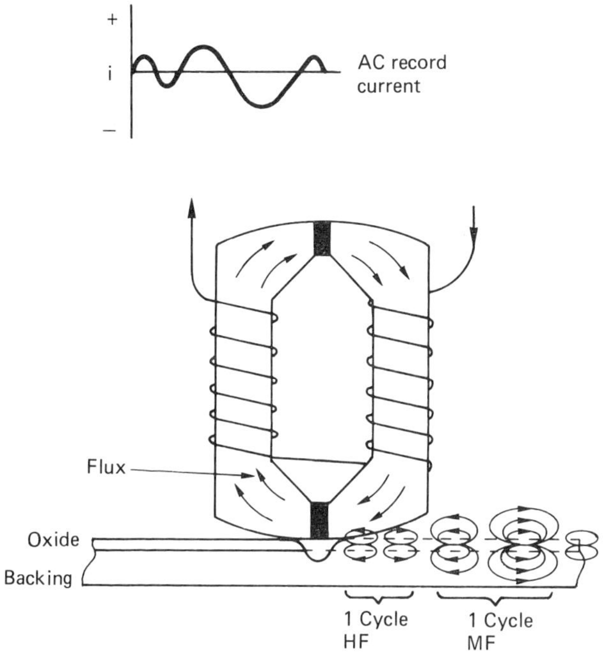


**Figure 8.1** Magnetic imprint on tape represents, to a greater or lesser degree, variations in current flow through the record head.

The addition of this bias signal results in partial demagnetization of the magnetic tape at high frequencies, where the recorded magnetic wavelengths are short, and this has traditionally been corrected by applying equalization in the recording amplifier.

## *The replay process*

During replay, the magnetized particles are caused to pass at a linear speed in front of a replay head, where their external flux links with a coil of wire, generating an EMF. The strength of this EMF is dependent on the rate at which the magnetic flux is changing, though this 'differentiation' of the flux strength is modified by the inductance/resistance of the head coil (Figure 8.3).

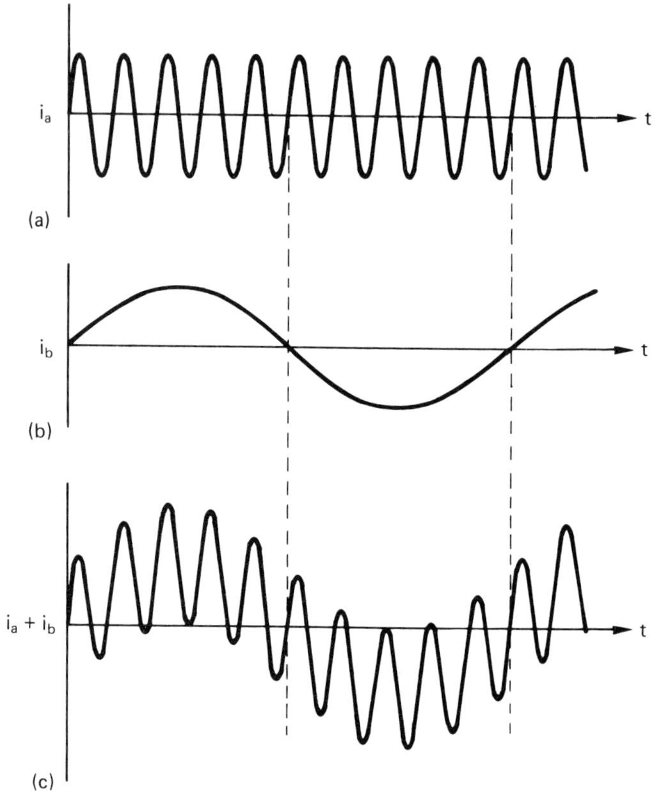


**Figure 8.2** High frequency bias (a) is a sine wave usually an order of magnitude higher in frequency than the highest audio frequency (b). In the recording head the audio and bias currents are superimposed. The process is one of straightforward addition (c). It is *not* modulation (despite the casual usage of the term 'mod' to describe the recording process).

As in the recording process, there are variations in the frequency response, mainly at high frequency, brought about by losses within the replay head, head contouring effects, and variations in the intimacy of tape-to-head contact. Again these variations have traditionally been compensated for by applying equalization during the replay process, using an alignment tape having known characteristics as a reference.

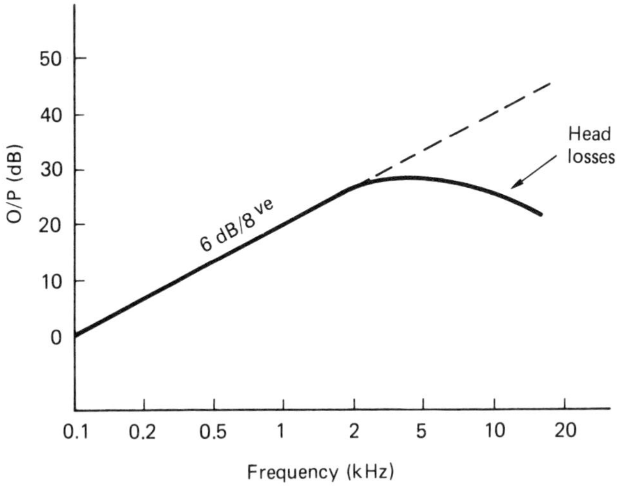


**Figure 8.3** At high frequencies tape heads suffer from a variety of effects causing loss of available output voltage.

## *Traditional record/replay shortcomings*

Apart from the non-linear relationship between recording current and corresponding magnetism on tape (which the application of high frequency bias does not completely eliminate – there is usually residual distortion in the order of 1–2%), there are other shortcomings in the record/replay process, which have been adequately overcome for audio recording, but which make traditional methods unsuitable for recording video. These shortcomings can be summarized as follows:

### *1. Variations in particle size*

Unavoidable with current techniques for manufacturing iron oxide powder, this variation in particle size results in noise which manifests itself in the form of hiss. This has been overcome by the application of noise reduction techniques (Dolby, dbx etc), and by the use of a high linear tape-to-head speed in order to ensure that the recorded magnets are as long as possible, and in any case, several orders of magnitude larger than the longest oxide particle at the highest recorded frequencies. (As an example of the foregoing, a frequency of

15 kHz, recorded at a linear tape speed of 38 cm/sec will result in magnet pairs of $1.2 \times 10^{-6}$ m length each (Figure 8.4). Iron oxide particles are in the order of $0.2\text{–}0.8 \times 10^{-6}$ m long.)

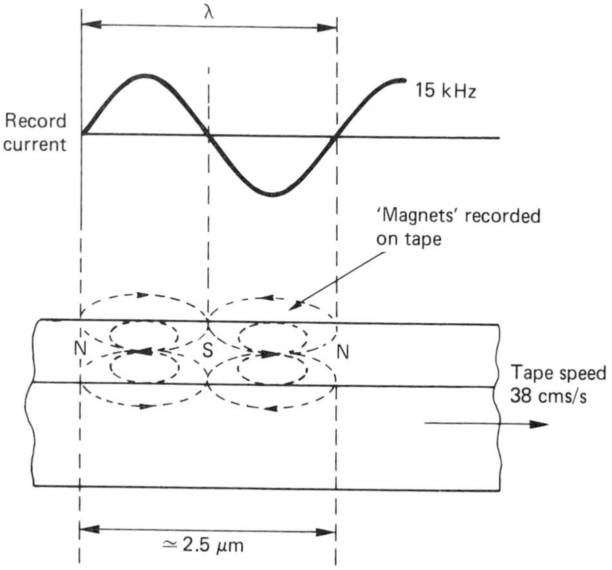


**Figure 8.4** Even at the higher audio frequencies, recorded wavelengths on tape are very short despite quite high longitudinal tape speeds.

### 2. *Variations in the oxide coating*

Variations in thickness result in corresponding variations in sensitivity. Should oxide particles clump together, or be absent (perhaps as a result of poor adhesion), then dropout will result. High linear tape speeds and wide recorded tracks have minimized dropout.

### 3. *Spacing and gap losses*

Lack of intimacy between tape and replay head in particular results in losses at high frequency (if the space between tape and head is one recorded wavelength, the signal loss will be over 54 dB, as Figure 8.5 shows).

The width of the replay head gap also affects the frequency response in the replay process. The output from the head rises from zero, at zero frequency, to a maximum when the replay head gap is one-half the length of the recorded wavelength, and

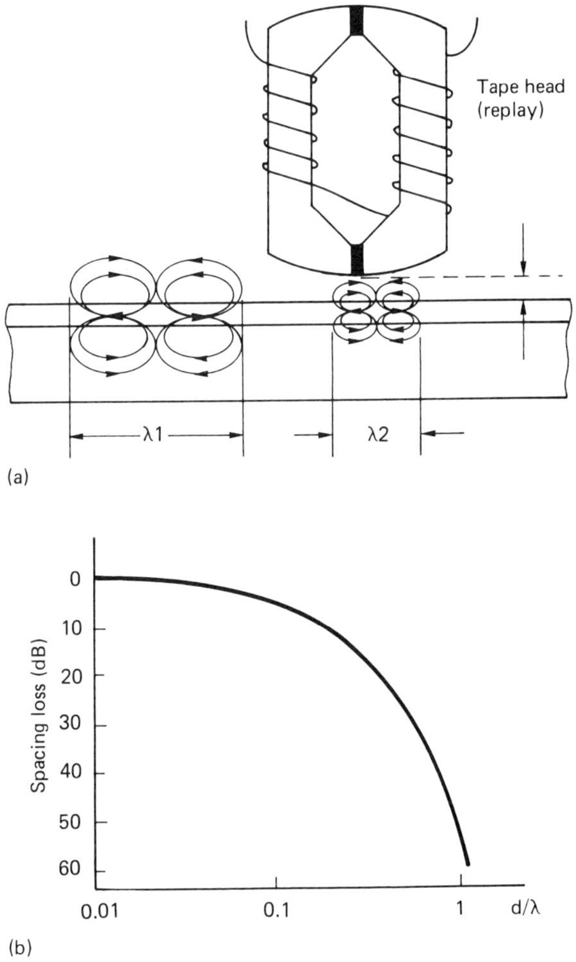


**Figure 8.5** At high frequencies recorded flux does not extend far from the surface of the oxide (a). If the tape is not in intimate contact with the head then flux linkage suffers. As soon as tape-to-head spacing with respect to wavelength (d/λ) exceeds a ratio of 0.1, significant losses occur (b). For video recording, where recorded wavelengths are in the order of 0.5–0.9 μm, the effective spacing between tape and head has to be less than 0.05 μm.

falls to zero again when the head gap is one complete wavelength long, this being brought about by equal and opposing magnet pairs within the head gap generating equal and opposite EMFs in the replay head coil (Figure 8.6).

Keeping high standards of maintenance, particularly cleanliness, combined with sufficient back-tension to hold the tape firmly against the head, helps ensure optimum high frequency response, as does making the replay head gap as small as possible, and a high linear tape speed.

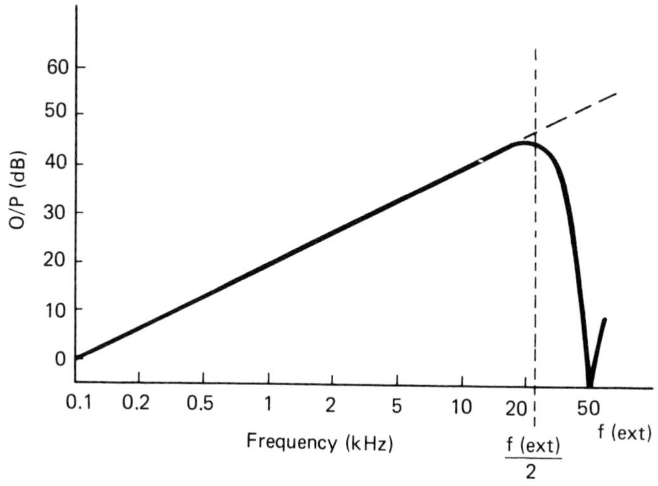


(a)

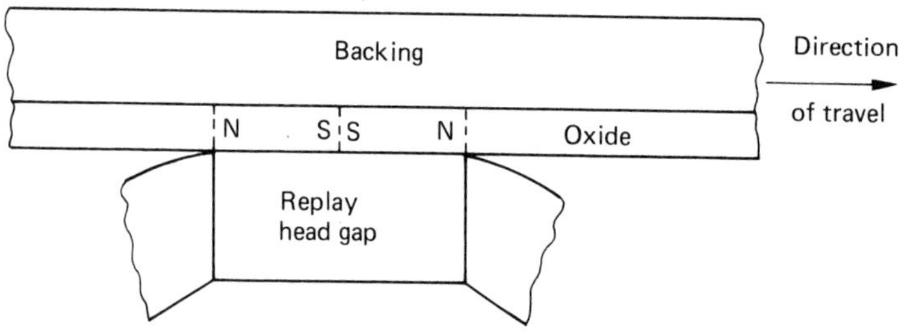


(b)

**Figure 8.6** All tape replay heads exhibit increasing high frequency loss leading to complete loss of output (extinction) at *f*(ext). The output from the head is greatest at *f*(ext)/2 (a). The cause of this extinction is the presence of mutually opposing pairs filling the effective gap in the replay head pole pieces (b).

*4. Bandwidth limitations*

The EMF generated on replay is dependent on the differentiation (rate of change) of the recorded flux. As a result, not only will the *O/P* rise with the recorded flux strength, but also with the frequency (replay head gap effects apart) as, for a given level of flux, the rate of flux change will double for every octave rise in frequency. For the audio signal, covering the frequency band from 30 Hz to 15 kHz (approximately 9 octaves), this will result in a difference of 54 dB in *O/P* level between lowest and highest recorded frequencies (Figure 8.7).

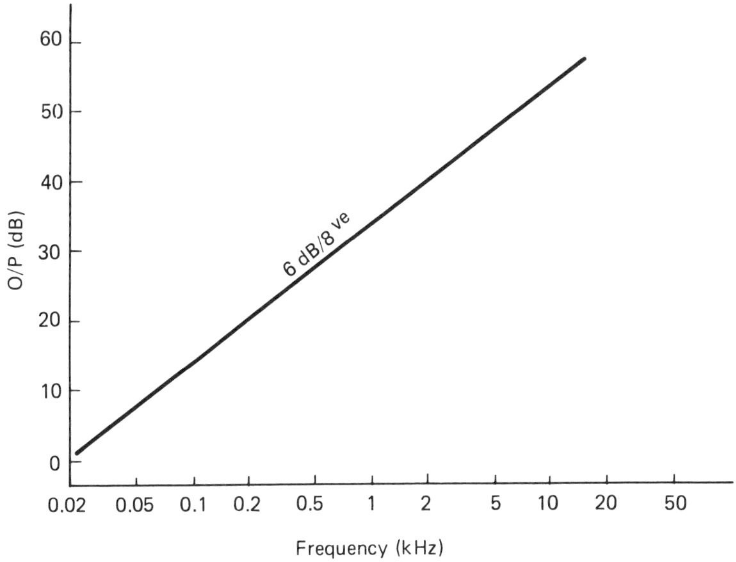


**Figure 8.7** Head and spacing losses apart, the output from a tape replay head will rise at 6 dB/8$^{ve}$.

Applying equalization to predetermined standards during both the recording and replay processes overcomes this problem, though the standards differ from country to country, and between tape speeds, making the interchange of programme tapes a process needing a great deal of care. Figure 8.8 illustrates the process.

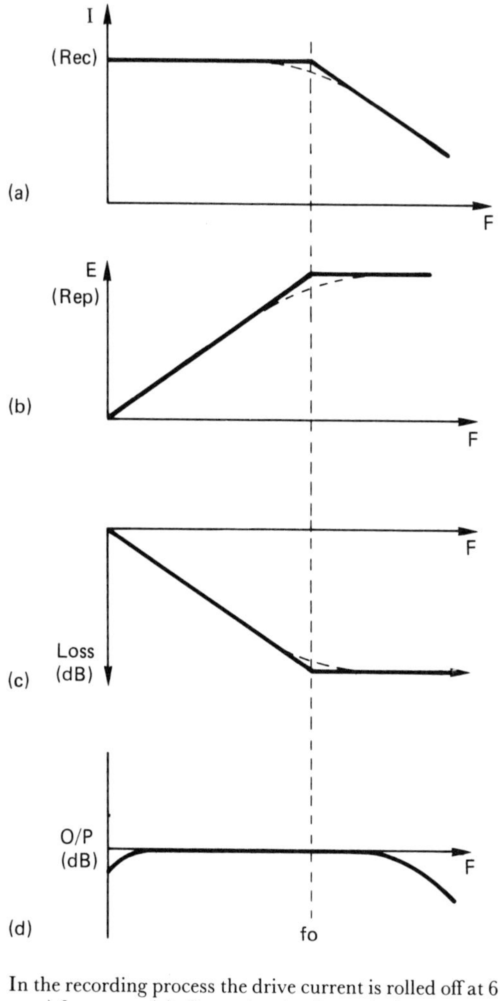


**Figure 8.8** In the recording process the drive current is rolled off at 6 dB/8$^{ve}$ above an agreed (turnover) frequency (a). On replay the theoretical output voltage will rise at 6 dB/8$^{ve}$ up to the turnover frequency and will then level off (b). Equalization on replay compensates for this rise in output voltage (c). The result is a frequency response that only requires compensating for low frequency and high frequency replay losses (d).

## *The need for improved oxides*

With the advent of cassetted audio tapes, an oxide formulation was required that would permit reasonable standards of tape hiss and frequency response at the low tape speeds employed, and that would have acceptable dropout characteristics with the narrow audio tracks necessary to squeeze four tracks, with guard bands, onto the width of audio compact cassettes.

The use of chromium dioxide provided the answer. Chromium dioxide particles can be manufactured with a much more even particle size than is possible with iron oxide, resulting in lower noise and a more extended high frequency response than would otherwise be possible. Coating consistency is also generally better as the particles are less easily magnetized than iron oxide, resulting in less dropout. The disadvantages are the need for stronger magnetic fields to both erase and record, and a worse tolerance to overload. The requirement for stronger erase and record fields is brought about by the higher coercivity of chromium dioxide (twice that of iron oxide), requiring twice the magnetizing force. The lower tolerance to overload is brought about by a hysteresis loop having a squarer form than that of iron oxide (which is generally said to 'bend gently into compression'). Chromium dioxide has a much greater linear portion than iron oxide, but goes quickly into saturation at either end, resulting in 'clipping' of the audio signal if it is too high in level (Figure 8.9).

In fact, chromium dioxide is not really suitable as a recording medium for component analogue video (CAV), since it does not satisfy the very stringent requirements needed to record the broad spectrum signal (in the order of 25 Hz – 5.5 MHz) at the slow write/read speeds employed. The development of metal particle tape, which can be polished (calendered) to a very smooth finish, and which has much more consistent magnetic characteristics, has been essential to the development of CAV.

## *Bandwidth requirements for video recording*

When considering the recording of video signals onto tape, account has to be taken of the much wider frequency spectrum required. For the 625-line PAL system used in the UK this extends from 25 Hz to 5.5 MHz, or approximately 17 octaves. The 102 dB difference in *O/P* level versus frequency that this implies is clearly unacceptable, so other means have to be found of recording the signal.

Even if the problem of coping with the 25 Hz to 5.5 MHz

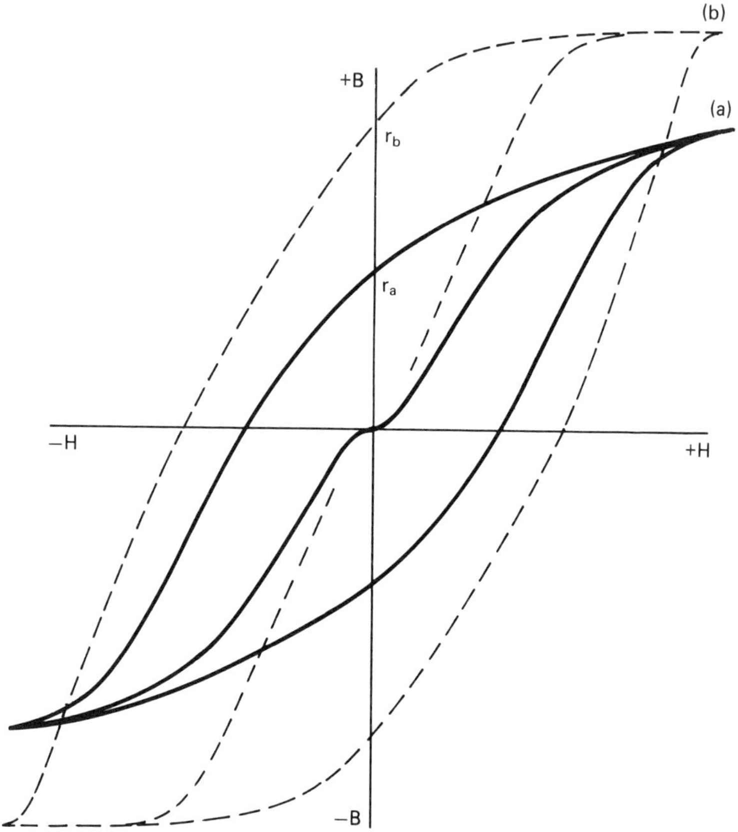


**Figure 8.9** The hysteresis loop for iron oxide (a) has a shape that curves gently into saturation but has a relatively low remanence ($r_a$). Chromium dioxide has a higher remanence ($r_b$), but its hysteresis loop goes into saturation much more quickly (b).

video bandwidth could be overcome, the size of the oxide particles, which have a length in the order of $0.5 \times 10^{-6}$ m, would require a tape-to-head speed of AT LEAST 5.5 m/sec to ensure the highest frequency to be recorded would produce a wavelength on tape of at least the length of two oxide particles. This would be extremely prodigal of tape, and indeed the operator would need to load a 20,000 m reel of tape for a 60-minute programme (assuming he could lift it!).

## *Use of frequency modulation for video recording*

In practice, the video signal is not recorded directly onto tape, but is used first to frequency modulate a constant amplitude carrier. It is this resulting constant amplitude signal that is recorded as a magnetic imprint on tape. This minimizes the variation in recorded levels that would otherwise occur with the 5.5 MHz wide bandwidth video signal. The use of frequency modulation also allows the DC component of the video signal to be recorded, as different signal levels have their corresponding modulation frequencies (Figure 8.10).

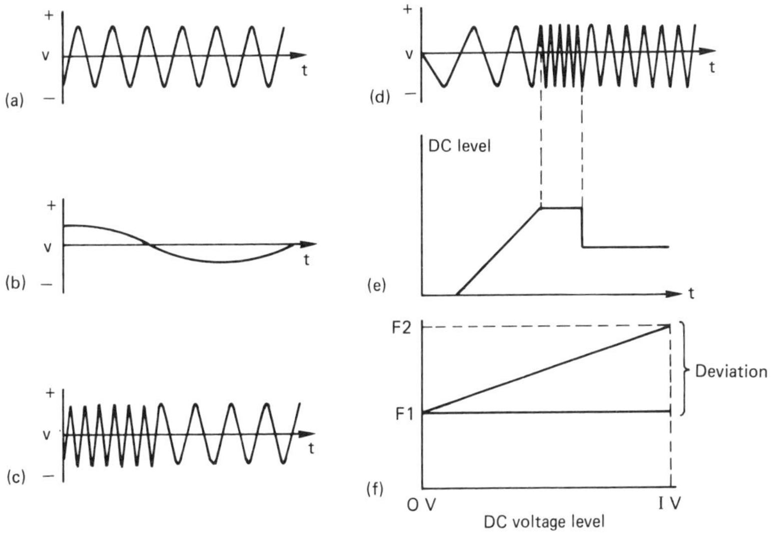


**Figure 8.10** Unmodulated FM carrier has constant amplitude and frequency (a). The frequency of this carrier is modified by a signal (b), the result being a change in frequency, but not amplitude, of the carrier (c). Specific carrier frequencies (d) can represent absolute DC modulation signal voltage levels (e). Sync tip, blanking and peak white video levels may thus be represented by an AC signal and so recorded onto tape. The difference between FM carrier frequencies representing sync tip and peak white is known as the deviation (f).

Yet another spin-off, where component analogue recorders are concerned, is that the luminance and chrominance components of the signal are recorded on physically different tracks on the tape, and there is space in the low bandwidth chrominance signal to allow two audio tracks to be FM recorded along with it, on the same set of tracks. This allows the recording of four audio channels in all, the traditional two longitudinal tracks, and two high quality FM audio tracks. The disadvantage is that the audio signals cannot be dubbed separately as, in effect, they form a part of the video recording (Figure 8.11).

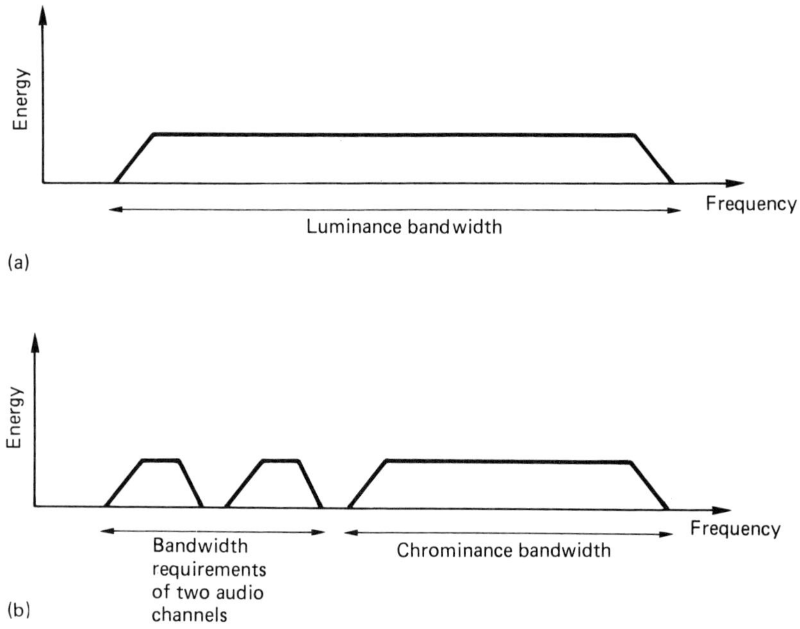


**Figure 8.11** In component analogue recording the luminance component is FM recorded (a). The bandwidth requirements are wide to permit the resolution of fine detail. The chrominance components (Cr, Cb) occupy a narrower bandwidth (b) as less fine detail is required. Audio signals can be recorded in the spare part of the spectrum.

## *Use of helical scan to improve read/write speeds*

A high enough tape-to-head speed is obtained by having the heads mounted in a drum which spins round, and wrapping the tape around the head in an open spiral. This results in a series of recorded tracks being laid in shallow diagonal lines across the tape. In this way, although the linear speed of the tape may be quite low, the writing speed will be very high. One typical system employs a linear tape speed of 0.066 m/sec, and has a writing speed of 6.9 m/sec, a writing to linear speed ratio in the order of 100:1 (Figure 8.12).

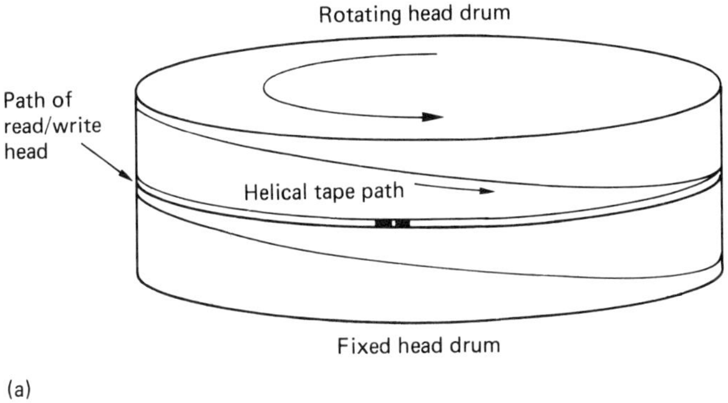


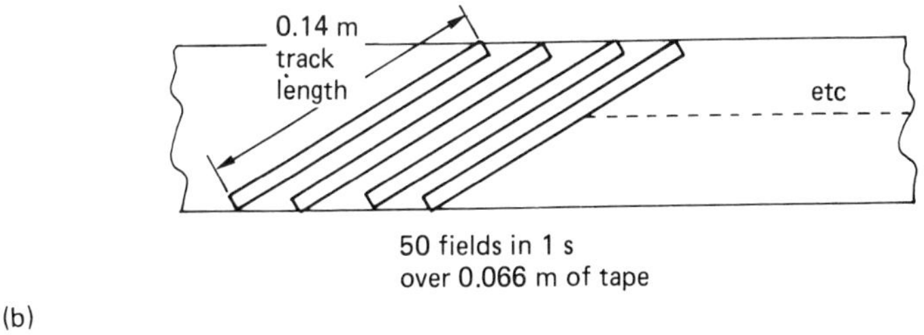


**Figure 8.12** The video tape is wrapped round a spinning drum in the form of a part helix (a). Heads set in this drum write and read diagonal tracks on the tape (b) as it moves at a relatively slow longitudinal speed through the transport system.

## *Implications for audio on CAV recorders*

There are several implications for the recording of audio onto any helical scan video format that result from the above considerations.

Firstly, the tape formulations used, chromium dioxide, and metal in the case of component analogue recorders, do not permit any latitude regarding over-modulation. This being so, many helical scan recorders, and in particular component analogue video (CAV) recorders, Betacam, Betacam SP, and M II, are fitted with limiters to prevent over-peaking. The limiters cannot always be switched out of action, much depending on the machine.

The result of over-modulation during recording is predictable; severe distortion if the limiter is switched out and, with the compressor/limiters switched in, a loss of dynamic coupled with an apparent increase in the level of ambience. The compressor/limiters on component recorders are not ganged, so an off-centre sound source that brought the limiters into action would not bring them both in to an equal extent and as a result background sounds would shift away from that side, returning as the limiter/s released.

To improve the audio signal/noise ratio on the longitudinal audio tracks, Dolby-C noise reduction is employed. This system applies 20 dB of compression to the signal to be recorded, at those frequencies to which the ear is most sensitive (Figure 8.13).

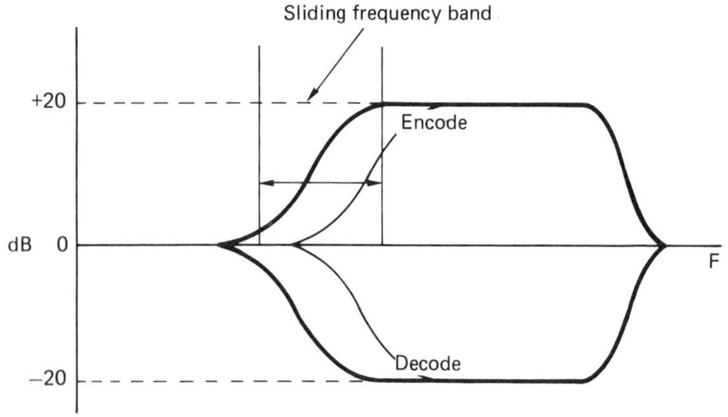


**Figure 8.13** Dolby-C applies up to 20 dB of compression upwards of low level audio signals prior to recording, and expansion downwards after replay. The frequency range affected varies according to the spectral energy distribution of the signal. For this reason the encoders and decoders are not linked between tracks.

It is important that a Dolby-C recorded tape is post-produced via a Dolby-C unit, and equally vital that a non-Dolby tape is not subjected to Dolby processing at a later stage when the signal is expanded by 20 dB. If either of these situations arises, the result will be an unbroadcastable recording. Again, correct logging of the road tapes will help avoid errors.

The audio tracks on CAV recorders are in the order of 0.4–0.6 mm wide, compared with the 1 mm width for the tracks on a 24-track, 2 in audio tape recorder. This need not be a problem as long as maintenance is up to standard. The layout of the tracks is shown in Figure 8.14.

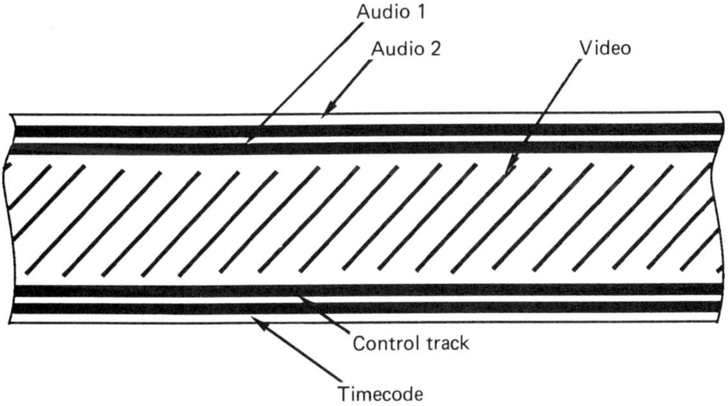


**Figure 8.14** CAV tape formats place the two longitudinal audio tracks alongside each other on the opposite side of the video tracks from the control and timecode tracks. Even so, the slow longitudinal tape speed can result in significant phase differences between these tracks.

To cope with the slow tape-to-head speed of the longitudinal tracks on CAV recorders, a large amount of equalization (pre-emphasis) is applied by the record amplifier, and there are indications that this can cause crosstalk between tracks at high modulation levels. Some types of machine may have a reduced high frequency response as an answer to the oscillation at high frequencies and high modulation levels brought about by the large amount of feedback provided.

## *FM sound tracks on CAV recorders*

For all the above reasons the longitudinal audio tracks on CAV recorders are generally thought of as unsatisfactory for good quality recording, and both Beta SP and MII offer high quality FM channels. Note the word 'channels'. FM sound on these recorders is incorporated as a part of the video signal for recording and replaying.

On CAV recorders the luminance and chrominance components of the video signal are separated. The chrominance information is matrixed into two colour difference signals, Cr and Cb. The luminance signal is used to frequency modulate a carrier, which is then recorded as a set of diagonal stripes on the tape. The two chrominance signals are time-division multiplexed together and this signal is then used to frequency modulate

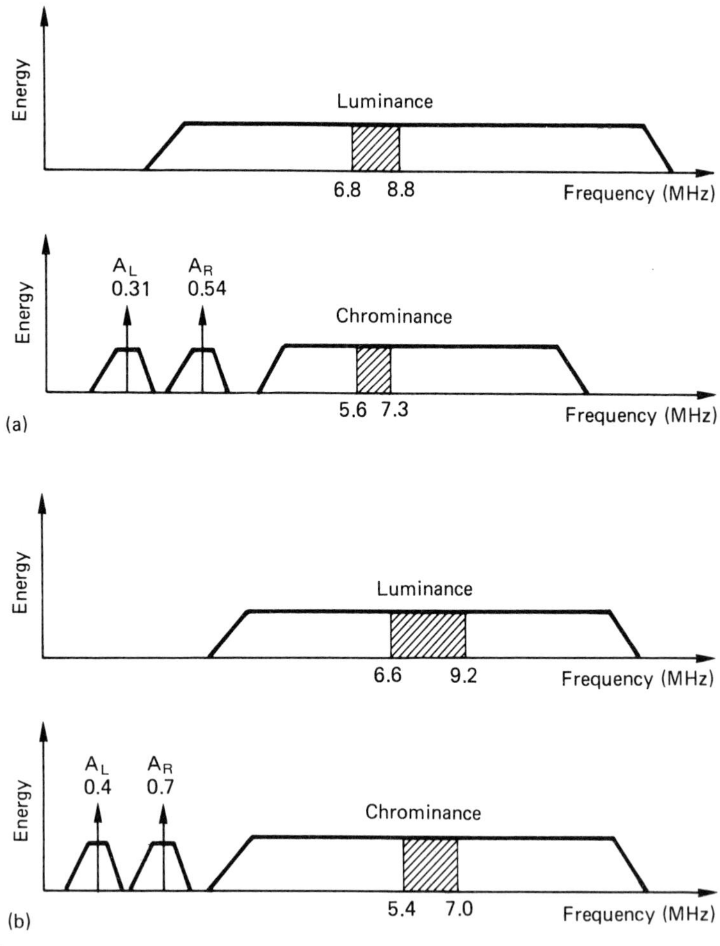


**Figure 8.15** Beta SP energy spectra for recorded luminance, colour difference and FM audio signals (a). Note the luminance and colour difference. FM carrier deviation frequencies are not central in the spectra (vestigial sideband). MII energy spectra (b) show equal upper and lower sideband distribution.

another carrier. The modulated chrominance signal can use a narrower bandwidth than the luminance carrier since chrominance signals require lower definition, the human eye resolving less detail in colour differences than in brightness differences. This narrower bandwidth leaves space for additional signals, and it is in this space that two frequency-modulated audio signals can be placed. They are not multiplexed but are recorded as discrete left and right signals on two different carriers. The carrier frequencies differ between the Beta SP and MII systems (Figure 8.15).

The resulting combined audio and video signal is recorded as a series of stripes on the tape, alternating with the luminance (Figure 8.16). As the audio signal is incorporated into the video prior to recording, later insert editing of video only or sound only is impossible, and the audio signal must be lifted off, thus losing a generation, prior to anything but the simplest post-production.

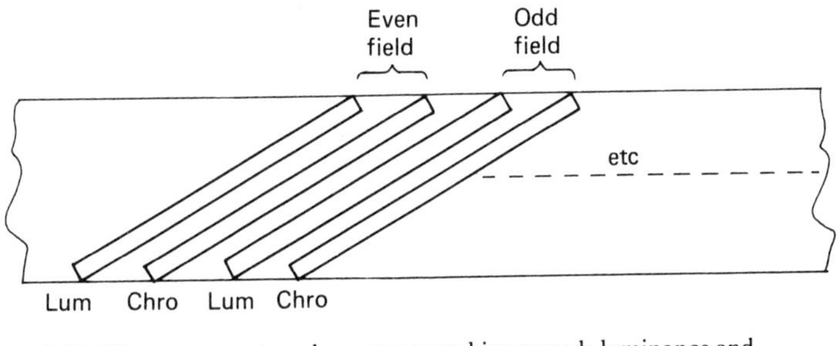


**Figure 8.16** The component analogue tape machine records luminance and chrominance on separate tracks. The FM audio signals are recorded along with the chrominance signal by the same head, so are usually referred to as 'channels' rather than tracks.

## *Digital sound on CAV recorders*

Digital (PCM) sound tracks are now available as an option. Compatibility with existing (non-digital) versions of the formats is obviously essential. This has been achieved by extending the wrap of the tape round the head drum, making for a longer written track, but at the same angle as non-digital versions, and extending the wrap in such a way that the extra track length will over-write longitudinal audio track 1 (Figure 8.17). Extra heads

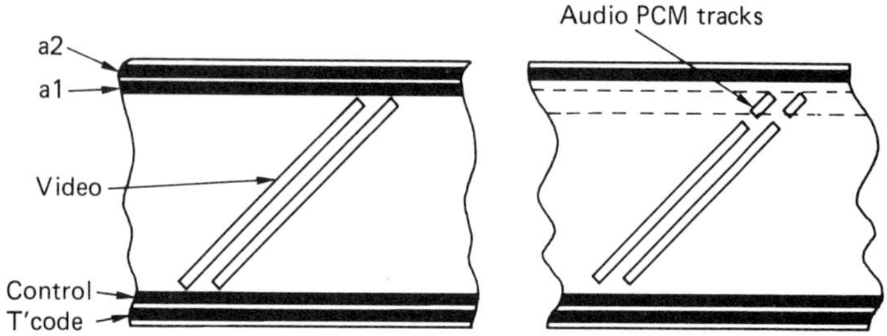


**Figure 8.17** Digital (PCM) audio tracks can be incorporated by extending the diagonal track length into the space previously occupied by one of the longitudinal audio tracks. This gives five audio channels.

are incorporated into the head drum to write and read two PCM tracks into this space.

As the audio is laid down discretely, it is possible to undertake separate insert editing of sound and vision during post-production.

## *Implications for the metering systems*

Frankly the problem of over-modulation on CAV recorders is not helped by manufacturers providing only VU meters on camcorders. What is really needed is a meter scale that provides linear needle deflections with linear dB level increases. The VU meter not only has a dB scale that is very compressed below −7 VU (corresponding usually, but not always, to −3 dBu), but has a 300 msec rise time. This can result in the VU meter under-reading signal peaks by as much as 13 dB on some types of programme material (Figure 8.18).

This property of under-reading signal peaks can mean that a VU meter needle flicking just up to the red sector (a nominal +4 dBu) just could be indicating signal peaks of up to +17 dBu. If the recorder has been set up as recommended by the manufacturer, tape overload could occur at +13 dBu I/P. In any case the limiter, if in circuit (which it usually is), will come into action at levels as low as +6 dBu depending on the precise format in use.

Because the compressors are fairly severe in action and are not linked, because the camcorder may well have VU meter

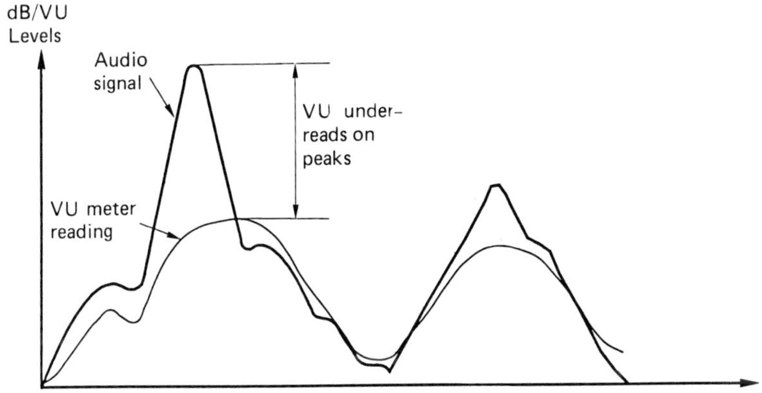


**Figure 8.18** The relatively slow reponse time of VU meters may cause them to under-read by up to 13 dB on some transient signals.

characteristics, and because levels are not easily controlled on the camcorder when shooting is underway, it is important that proper monitoring and control of levels is undertaken with a separate mixer, with operationally friendly meters, monitoring facilities, and gain controls.

# 9

# Field recorder formats

## *Sony U-Matic*

Sony's standard U-Matic format has two, low quality, sound tracks. The SP version does have Dolby-C but the quality is still barely adequate for stereo.

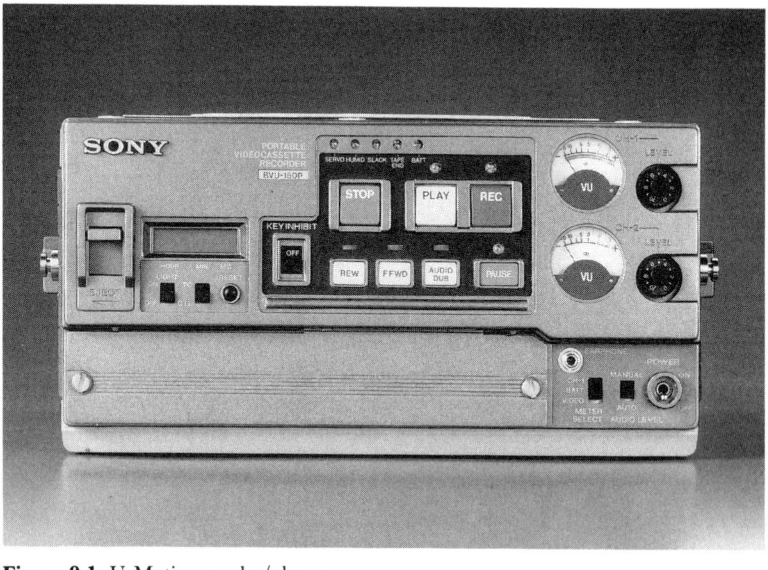


**Figure 9.1** U-Matic recorder/player.

### *Sony Betacam*

Sony replaced U-Matic with Betacam. It still only has two longitudinal tracks, but does offer Dolby-C noise reduction. Although not tolerant of multigeneration recordings, the quality is acceptable for mastering stereo in the field. But beware, as these machines are not stereo, but twin track, so no guarantees are given as to the phase correlation. This also applies to the longitudinal tracks of SP machines.

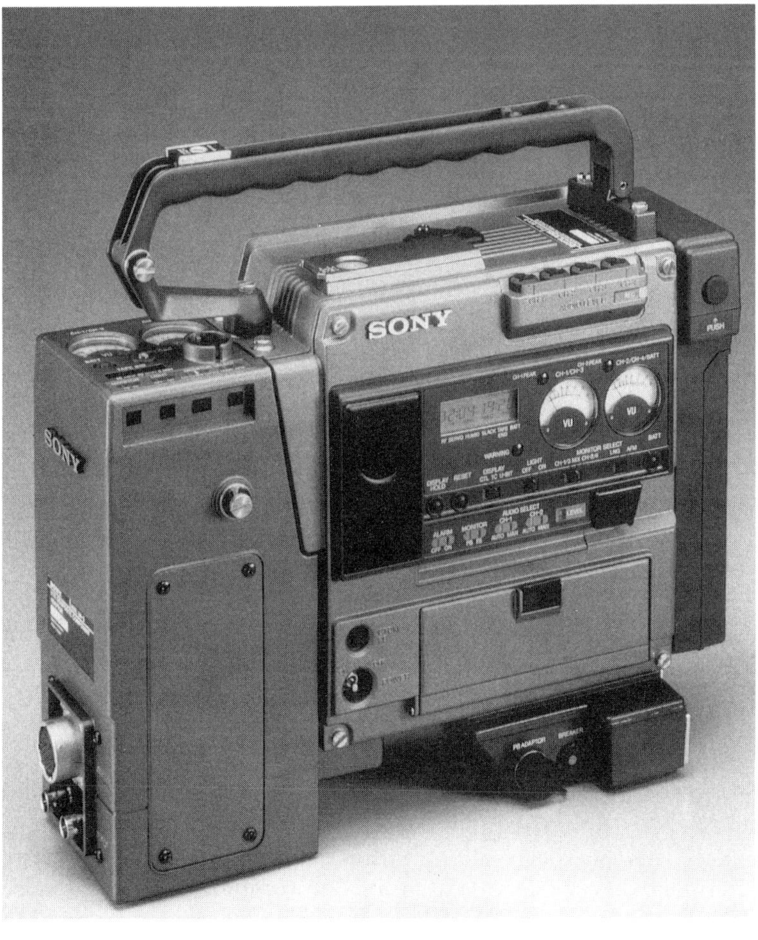


**Figure 9.2** Dockable Betacam SP recorder.

### *Sony Betacam superior performance (SP) and Panasonic MII*

Both Betacam SP and MII use metal tapes and offer high quality component video recording. Betacam SP machines, however, can record and replay standard Betacam oxide tapes. Each format has two longitudinal sound tracks using Dolby-C noise reduction, and two frequency-modulated (FM) sound tracks (only when using metal tapes) that are embedded in the video colour signal. Therefore audio dubbing on the FM tracks is not possible.

It is practical to use the longitudinal tracks of either format for stereo recordings but the FM tracks are far superior. Possibly their greatest asset is the frequency multiplexing of the two signals as this eliminates interchannel phase errors.

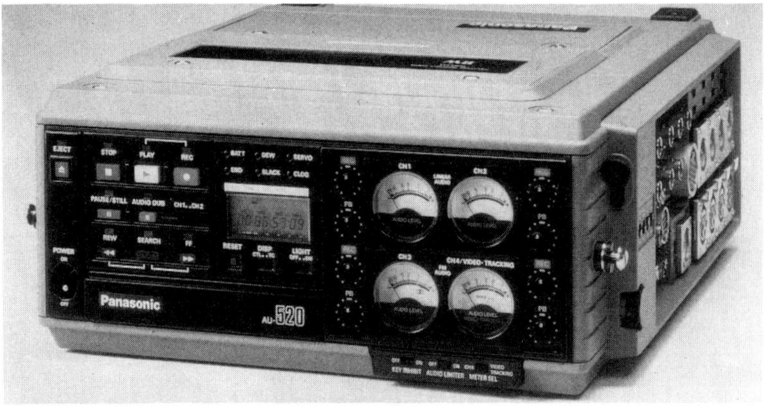


**Figure 9.3** Panasonic portable MII recorder/player.

Although there are four audio tracks available these are not 'four-track' formats. There is no accurate or stable phase correlation between the longitudinal and the FM tracks. They should be thought of as 2 × 2 track recorders. The Sony range of one-piece camcorders in fact only have two inputs, arranged so that the same signal is recorded on channel 3 as channel 1, and channel 4 is similarly linked with channel 2. It is critical that the phase between the two longitudinal tracks is optimized. There are internal controls available for fine tuning the relative phase of tracks 1 and 2 but this should only be attempted by trained engineers.

These phase discrepancies are not only a problem with AB recordings but will also do great damage to MS recordings with often minimal interchannel errors having a major effect on the stereo image. In absolute terms, MS tolerances need to be far tighter than AB. Both the formats offer machines with similar facilities.

The currently available camcorders do not have stereo monitoring; you cannot feed the auxiliary/return input of a mixer with a stereo signal. Instead they offer a mono output, switchable between input, and 'confidence playback', whilst recording. Each system has this facility, giving low-quality, off-tape monitoring of video and the longitudinal sound tracks during recordings. MII confidence does, however, seem to be of a higher quality than Betacam SP. They both suffer from timecode breakthrough, but this can be reassuring, as it confirms that code is being recorded.

One of the major problems with all the camcorders is their audio level controls. It is far too easy to knock these from their line-up settings. Although the MII equipment is certainly better in this respect, some form of lock is still needed. Note that camcorders have an additional level control on the camera body, so that the camera operator can adjust the sound level. On the MII AU400 this affects only the camera microphone, while in Sony equipment it controls channel 1. Again it is possible for this to be inadvertently moved. It is therefore absolutely vital that line-up is repeatedly checked throughout the day.

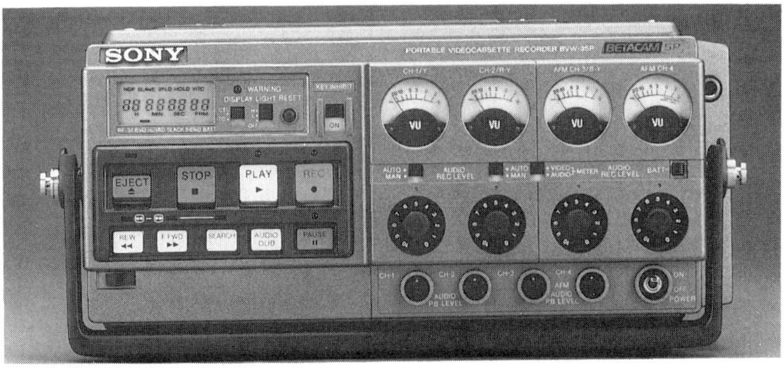


**Figure 9.4** Sony portable Betacam SP recorder/player.

The large portable player/recorders are equipped with better meters than the camcorders, but although both formats are fitted with stereo headphone sockets, only the MII machine, the AU500, has a stereo headphone output.

Both SP and MII have a digital sound option, but only on the studio machines. At the time of writing there are no plans to provide this on the portables.

### Dolby-C

When using either Betcam SP or MII for stereo recording, it is necessary to be aware of their limitations. Dolby-C is a powerful noise reduction system and all the machines use it. However, it is very critical of the frequency response of the recorder. Unless you have a transparent machine, the decoder will mis-track, causing unpleasant audio pumping. Because of the tape formulations, more designed for video signal than audio, and the relatively slow longitudinal head-to-tape speed, the frequency response of the longitudinal tracks is rarely stable. This is acceptable in mono but in stereo causes image shifts.

Both formats have switchable Dolby processors on the longitudinal tracks but always use noise reduction on the FM tracks. Be aware that the Dolby switch on the MII equipment is not easily accessible. On the AU400 it is between the camera body and the VCR, while on the AU500 it is mounted on an internal PCB. There is an indicator light to show when the Dolby is active.

### FM tracks

The specifications of the frequency-modulated tracks are impressive, but the system does have its limitations. You are left in no doubt of this when you over-drive the FM tracks, as they distort almost immediately, and MII is especially prone to this. In addition, the bandwidth of the FM signal is proportional to the audio level and therefore has to be tightly controlled to avoid interference with the other signals being processed. Both formats use powerful limiters to hold audio signals within the specified limits. These are in addition to the switchable limiters and are always in circuit. They are protection limiters and signals should normally be kept well below their threshold.

As the FM signals are processed as part of the video signal, any picture disturbance may well cause drop-outs on sound. Because it is a multiplexed signal, these drop-outs will be heard across both tracks and are much more obvious and offensive than their longitudinal track counterparts. Tapes that have been passed through the machine only a small number of times quickly start to show signs of drop-out and should be avoided. The FM tracks do not provide a confidence playback. They remain in the electronic-to-electronic (E-to-E) mode, so tape drop-outs can go undetected.

### *Limiters on video machines*

For reasons already explained, manufacturers have included permanent overload limiters in the FM circuits. Additional input limiters for all the audio tracks are also fitted, although they are usually either all on or all off. The Sony BVP200/300/400 have small circuit board switches inside the left-hand side cover that select limiters on channel 1 and/or 2.

There is no way of ganging either the protection or input limiters, so it is advisable to avoid using these devices for stereo recording. External stereo limiters should be used so that the protection limiters are never activated.

### *AGCs on video machines*

Automatic gain controls are crude devices for controlling sound levels. Their use in any professional sound recording should be avoided. MII machines do not have an AGC facility.

### *Meters on video machines*

All the machines come equipped with a selection of meters. There is one per audio track on the portable machines, and two, switchable between tracks 1/2 and 3/4, on the camcorders. As with all stereo, separate meters are inadequate for anything other than line-up. Some Betacam machines, which have peak reading bargraph meters, do however give some interchannel information. None however are switchable to MS metering. In most situations the meters on the recorder are only used for line-up so perhaps they do not need to be as comprehensive as those on the mixer.

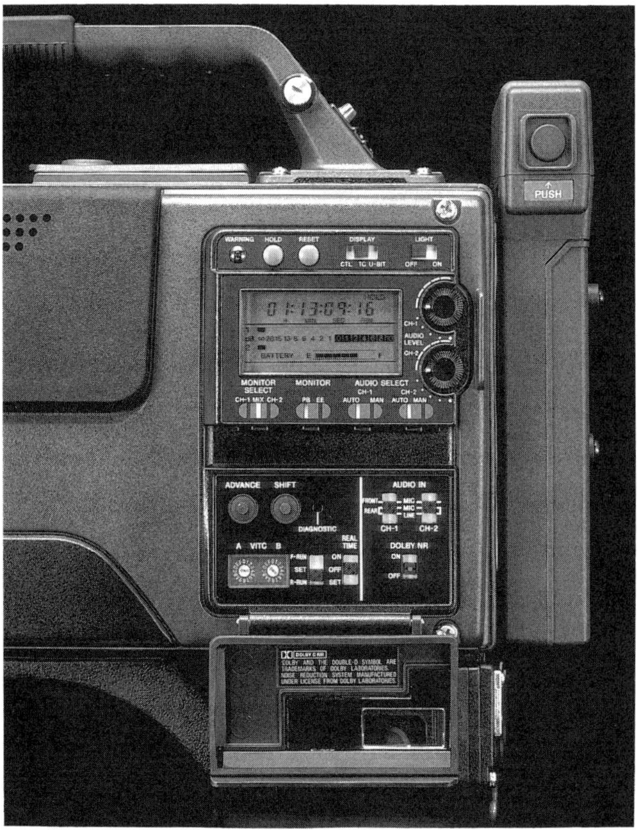


**Figure 9.5** Camcorder with dual bargraph meters.

A number of the camcorders have an audio meter in the viewfinder. This is a very simple mono meter. It really only gives an indication of the presence of an audio input. In Sony equipment this only shows the level on channel 1, while on MII it follows the audio selecttor switches.

## *Sound on digital video formats*

At present there are three forms of digital video. D1 is a component system while D2 and the half-inch Panasonic D3 format are composite. The technicalities of the formats need not concern us here. For the recordist they all offer four independently addressable tracks of digital sound and a single analogue guide track. Timecode is recorded on a longitudinal track and as vertical interval timecode (VITC).

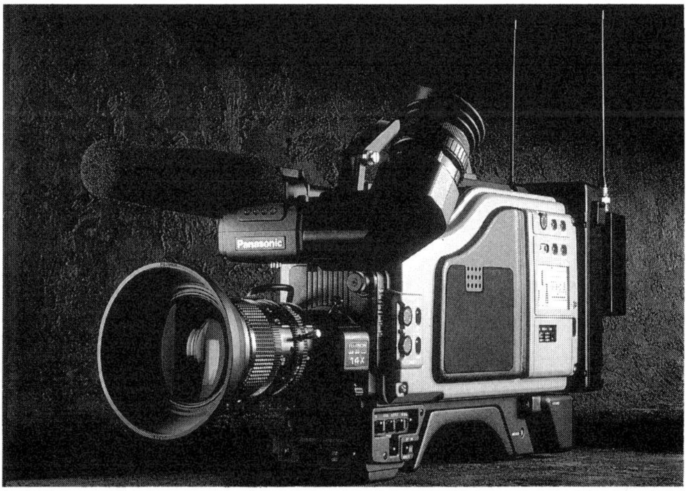


**Figure 9.6** Typical camcorder.

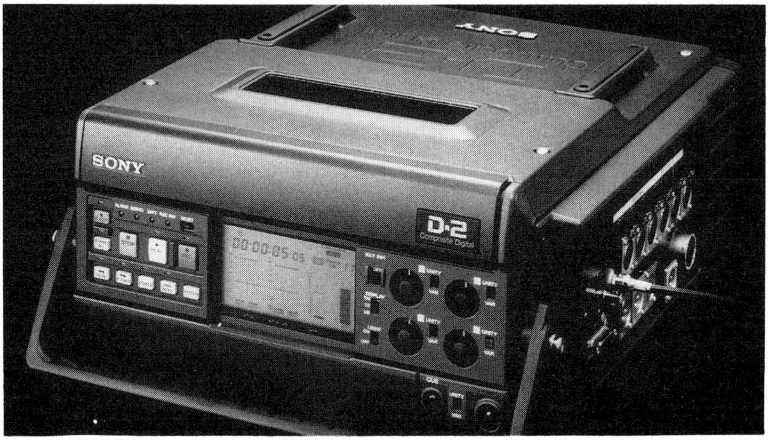


**Figure 9.7** Portable analogue recorder with integral timecode generator.

## *Stereo sound on Video 8*

This format has two options. Option one gives pictures and stereo digital audio, time compressed 10 bit at a sampling rate of 31.25 kHz. Option two uses the whole width of the tape to record six stereo signals. The six stereo tracks cannot be recorded or replayed together; you are only increasing the storage density, it is not a multitrack recorder. This is because, unlike other formats that use a control track to align the video head, Video 8 uses signals encoded in the recording. So different segments have no fixed relationship in time.

### *Sony Hi8*

Not originally designed as a professional format, Video 8 has recently come of age with the introduction of the new generation of Hi8 Semi Pro machines. With stereo FM tracks or a mono FM track and two PCM digital tracks, depending on the machine, Hi8 can produce better quality sound than many of the so-called professional formats. The format supports a timecode signal recorded between the video and the PCM tracks. It is separately addressable so tapes can be post-striped if required.

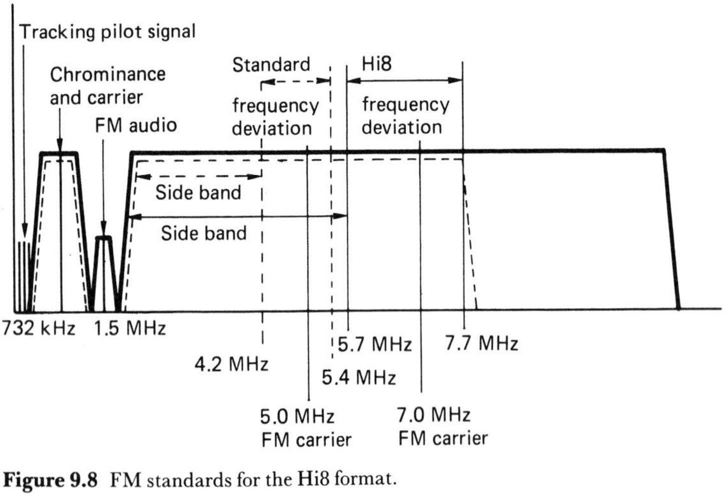


**Figure 9.8** FM standards for the Hi8 format.

### *S-VHS*

The VHS system has been around for many years and is firmly established both as a consumer and industrial standard. Recent improvements have allowed broadcasters to introduce the format into some ENG situations and off-line edit suites. S-VHS follows the current trend of having two longitudinal tracks and two FM tracks. The low tape speed means that the longitudinal tracks are not suitable for stereo. The frequency modulated (FM) tracks are, however, infinitely better quality and cope well with stereo signals, although incorrect video tracking on playback will cause dramatic drop-outs on the sound track.

## Separate audio recorders

However high the quality of the video sound tracks becomes it will always be preferable for the sound recordist to be independent of the camera. A radio link giving a guide track on the video will assist editing, but is not reliable enough for high quality recordings, especially in stereo. What is required is a portable stereo audio recorder, either analogue or digital, with a timecode generator. Whether to jam synchronize the recorder's timecode generator to the video recorder's generator or allow the dubbing mixer to synchronize the sound and pictures is a decision to be made in conjunction with the post-production personnel.

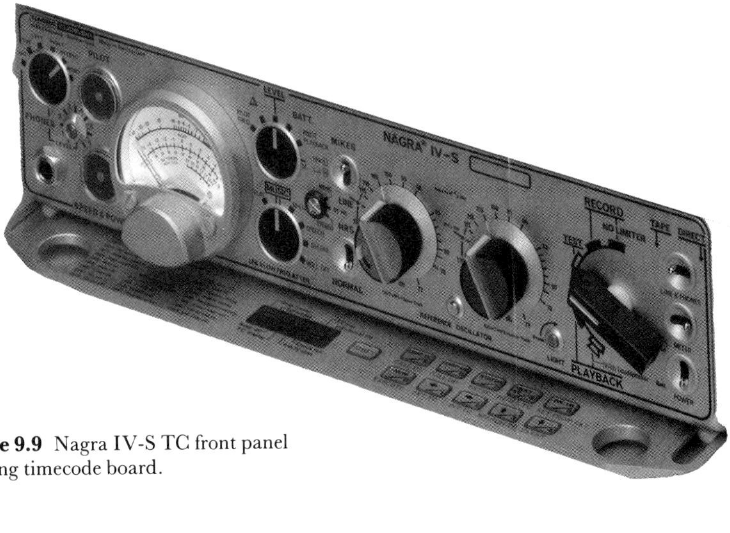


**Figure 9.9** Nagra IV-S TC front panel showing timecode board.

Although a radio link is not recommended for stereo sound it is a perfect way of sending a timecode signal from the camcorder to the audio recorder. This is then used to constantly update the internal timecode generator of the recorder and is usually very tolerant of any spurious noises produced by the radio link.

### *Open reel stereo tape machines*

As with any equipment used for stereo, the line-up of the tape machine is far more critical than in mono. Interchannel errors must be minimized. Today's machines offer very high electronic specifications but all location equipment suffers from mechanical shocks. Therefore head azimuth should be checked and re-checked at the start of every day. A test tape with 1 kHz at 0 dB and 10 kHz at −10 dB should become part of every sound kit. Used in conjunction with an S meter that has a +20 dB switch, quick and easy alignment is possible. At 1 kHz you need at least 20 dB separation between the M and the S, while at 10 kHz use the S+20 dB and look for at least 12 dB separation. All stereo recordings should be made at 15 ips to minimize any errors that remain.

### *Digital audio on video*

#### *Sony PCM 1610/30*

Although now superseded by other digital formats, this technique was the first to offer digital quality sound at a reasonable price. Using a PCM adaptor, an analogue audio signal is converted to digital form, then transcoded to a pseudo-video signal that is recorded on a U-Matic VCR that can have its drop-out compensator disabled. Using a 525/60 monochrome VCR, this system became the international standard for compact disc mastering.

#### *Sony PCM-F1*

The pseudo-video technique was taken a step further with the introduction of the Sony PCM-F1. Built as a low-priced consumer product it rapidly turned up in recording studios across the world. Although a fine machine, it does not offer direct digital inputs and outputs. There is only one D/A converter that is switched between the two output signals. Some machines were modified for digital-in digital-out, but this format has been overtaken by both DASH and DAT machines.

### *Digital audio tape (DAT)*

DAT machines offer many advantages to the location stereo recordist. Their 16 bit 48 kHz sampling gives excellent quality. Because it uses a multiplex system, the chances of interchannel errors are effectively eliminated. Many of the reservations about recording MS are therefore withdrawn. The physical size of most of the units makes them easy to carry and the relatively cheap tape costs should make them attractive to the broadcaster. Until recently, none offered reliable location recording of timecode. The Sony 2000 does have a timecode facility, but it suffers from drop-outs if the machine is moved. However, most do have the facility to record a 'start ID' in the data sub-code which can be used for later synchronization, as the machines have an extraordinarily stable playback. So much so, that free running synchronization is possible for some minutes!

The latest machine from Stellavox, the Stelladat, and the PD2 from Fostex are the first portable DAT machines to offer full SMPTE/EBU timecode, embedded in the data stream. With these machines, digital stereo sound for television location recording has become a reality.

There are studio players that have full timecode, and because you are working in the digital domain there is no loss of quality dubbing the master to a timecoded tape. Whether you choose to use a pre-striped tape or match to the video master's code will be decided by the time available and the in-house preferences.

One technique that attempts to give you some chance at timecode synchronization, if you don't have a Stelladat, uses the audio tracks to record about 20 seconds of timecode from the video recorder/camcorder at the start of a recording. The input is then switched back to audio. In the dubbing suite this burst of code is used to jam-sync a timecode generator, whose output is dubbed with the audio onto a syncable timecode DAT machine.

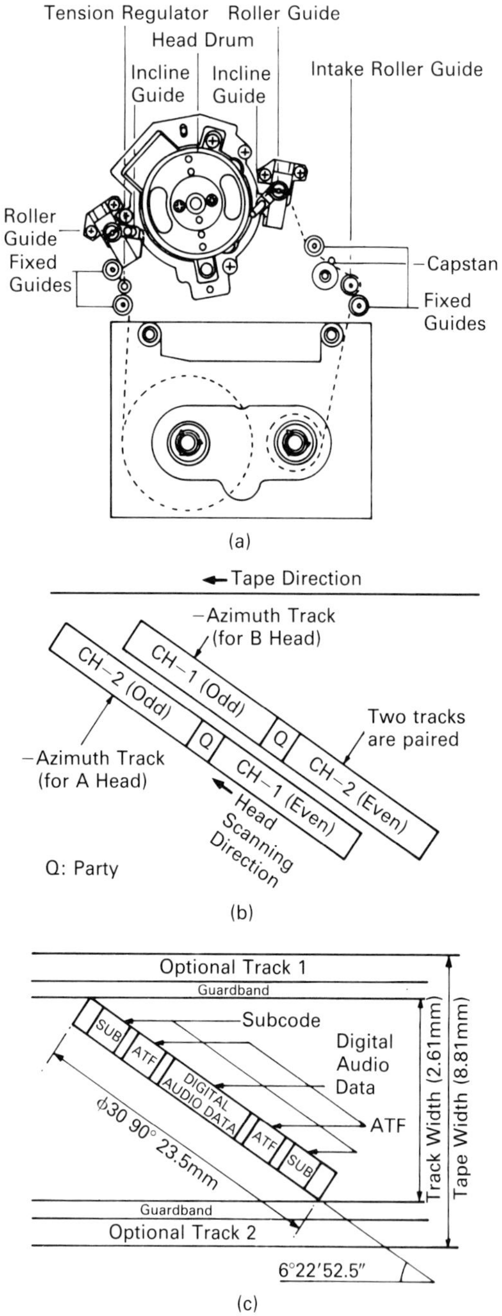

Tension Regulator
Roller Guide
Head Drum
Incline Guide
Incline Guide
Intake Roller Guide
Roller Guide
Fixed Guides
Capstan
Fixed Guides
(a)
Tape Direction
Azimuth Track (for B Head)
CH-1 (Odd)
CH-2 (Odd)
Q
Q
Two tracks are paired
Azimuth Track (for A Head)
CH-1 (Even)
CH-2 (Even)
Head Scanning Direction
Q: Party
(b)
Optional Track 1
Guardband
Subcode
Digital Audio Data
ATF
SUB
ATF
DIGITAL AUDIO DATA
ATF
SUB
φ30 90° 23.5mm
Track Width (2.61mm)
Tape Width (8.81mm)
Guardband
Optional Track 2
6°22′52.5″
(c)


**Figure 9.10**

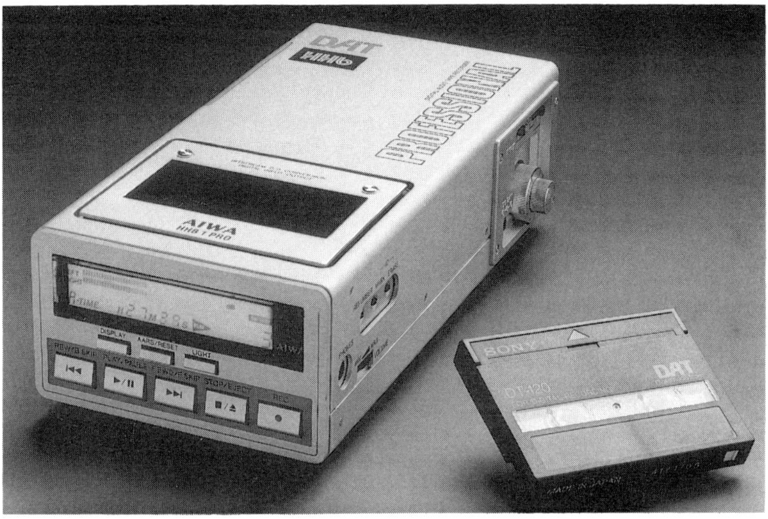


**Figure 9.11** Portable DAT machine.

Using a DAT edit suite, one could use a video edit list (EDL) to compile two DAT masters, incorporating pre-programmed audio overlaps. The locked DAT players then give you the equivalent of four track playback.

Despite the attraction of DAT, it does have some limitations. The format was developed for the consumer, rather than the professional, market, which is why it is only 16 bit, two track. Many believe that a mastering format needs to be around 20 bit. Stereo production requires the ability to access four audio tracks, rather than two. Locking together two DAT machines is an option, but it is not ideal.

## Nagra Digital open reel tape machine

The Swiss manufacturer Nagra has dominated the film sound market for decades. They realized that they could use their knowledge gained in data recorders to produce a digital audio machine. Rather than deciding what they could build, they started with a set of goals. These subsequently led to an impressive machine and an international standard.

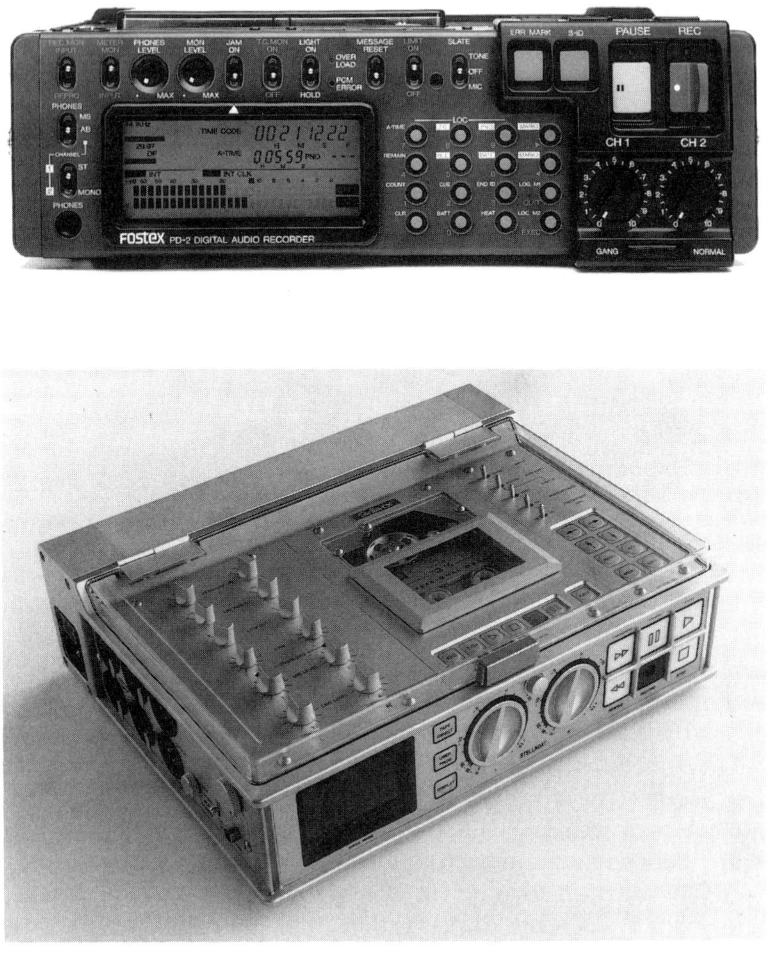


**Figure 9.12** Fostex and Stelladat Broadcast DAT machines.

The Nagra Digital offers four tracks each of 24 bits, 20 bit audio plus 4 data bits; full AES-EBU inputs and outputs; three hours' recording time with full confidence playback; standard longitudinal timecode readable up to full shuttle speeds; and one analogue cue track. Sampling rates of 48, 44.1 and 32 kHz are available.

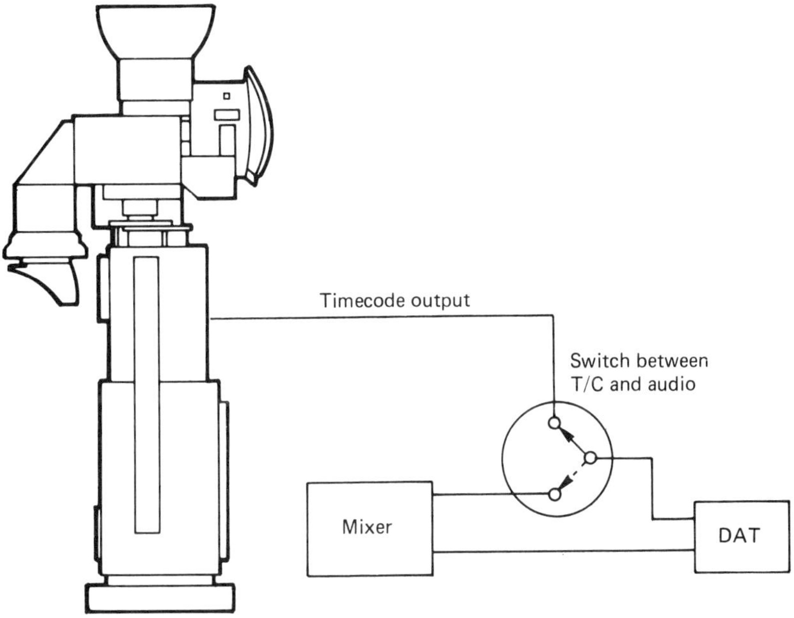


**Figure 9.13** Approximately 20 s of timecode is recorded at the start of each take to allow separate synchronization.

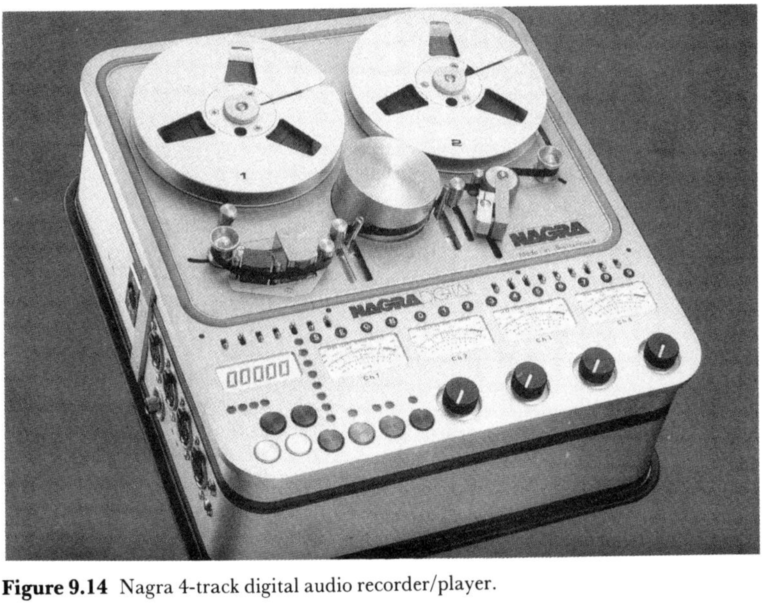

**Figure 9.14** Nagra 4-track digital audio recorder/player.

Using their digital experience to the full, Nagra have included a read-before-write option. A large amount of RAM is used to make non-destructive edits, using the instruction 'skip from here' and 'end of skip' in conjunction with very rapid spool motors. This large RAM capability also allows the format to tolerate physical splice edits. The major limitation of the machine is its size and weight. It is best considered transportable rather than portable.

## *Multitrack recorders*

There will be times when circumstances dictate that the individual microphone mix and stereo width would be best left to the dubbing mixer. If you have no idea how your element is to be edited into a programme, it may be advisable to leave as many options open as possible. There are sometimes just too many limitations in location equipment, especially the monitoring, for the recordist to judge the balance accurately. At these times, a multitrack recorder is a very useful tool.

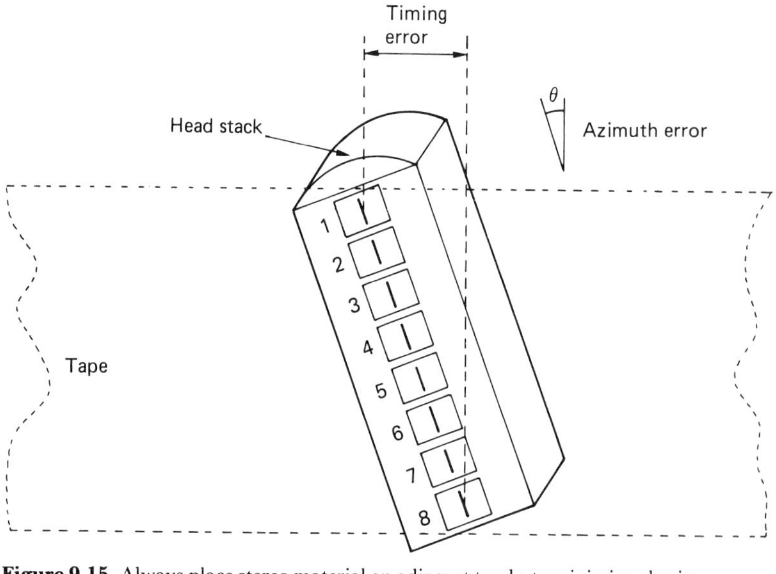


**Figure 9.15** Always place stereo material on adjacent tracks to minimize phasing caused by azimuth errors.

You will find that stereo working rapidly uses up a considerable number of tracks so it is advisable to plan your desk groupings and track configurations carefully.

Always keep stereo pairs on adjacent tracks, thereby minimizing the effect of any machine azimuth errors. Don't record in MS unless you have a good reason to, as the electrical and mechanical tolerances of the multitrack will affect them more than their AB equivalents.

Try not to mix AB and MS recordings on the same tape or programme as this may cause confusion later.

Sync heads do suffer from crosstalk more than the playback heads, so where possible avoid their use for stereo.

## Stereo and video editing

Any system is only as good as its weakest link. At present the video editing process is simply not up to dealing with stereo sound. Unless you have digital video machines that offer four tracks of PCM sound you cannot easily create overlap or chequerboard sound edits.

### *The BEL 6000*

A device like the BEL BDE 6000S is a simple and effective way of getting stereo mixing, without actually having four tracks! Instead you have a two-track solid state digital recorder with timecode that can record about 45 seconds of stereo sound.

The audio from the video master replay machine, from perhaps 5 seconds before the edit point to 5 seconds beyond it, is dumped to the digital recorder whilst the video is transferred to the master record VCR. Timecode is taken from the master record VCR. A sound 'in-point' is chosen prior to the video edit and the edit is started.

The BDE 6000S replays as it receives the identical timecode from the record VCR and will lock up in around 30 microseconds. At the sound 'in-point' you have the digital recorded sound faded up and you crossfade, at the appropriate point, to the sound from the replay machine.

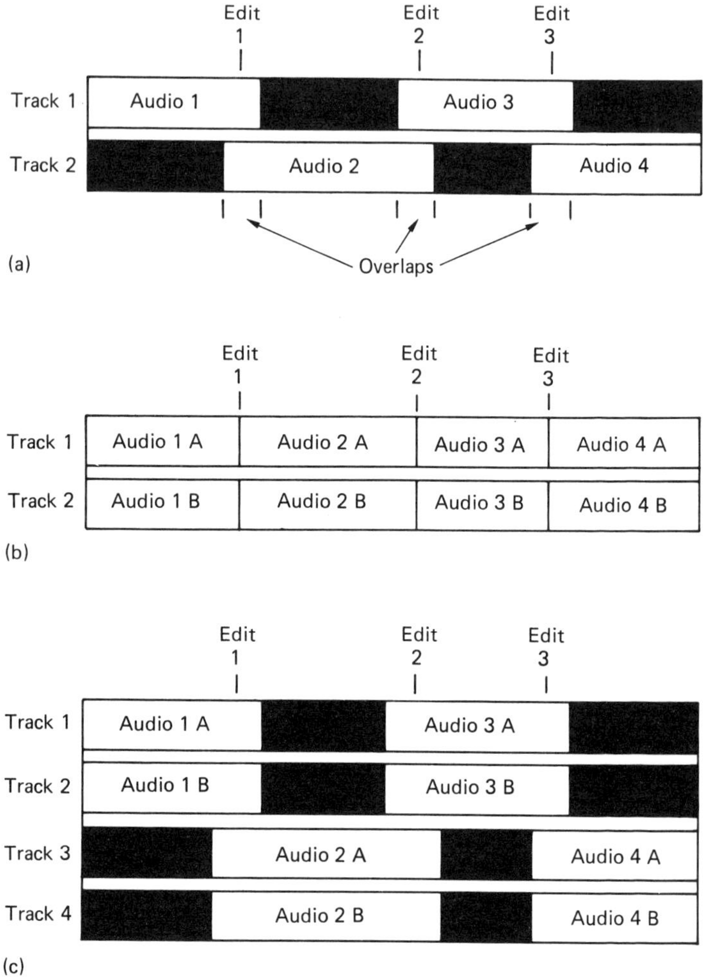


**Figure 9.16** Overlaps allow mono crossfades at video edit points (a). Stereo overlaps are impossible on two track recorders (b). Stereo overlaps need a fair track capability (c).

Thanks to an RS422 port on the BEL 6000, all this can be controlled from the video edit controller. The latest versions have a greater recording time and the ability to create segments in the memory.

This could all be done with a synchronizable tape machine, such as the Nagra T, but the simplicity of the digital solution makes it perfect for the video edit suite.

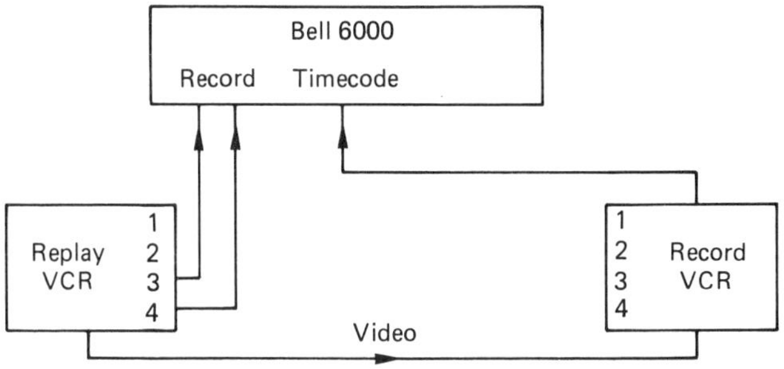


(a)

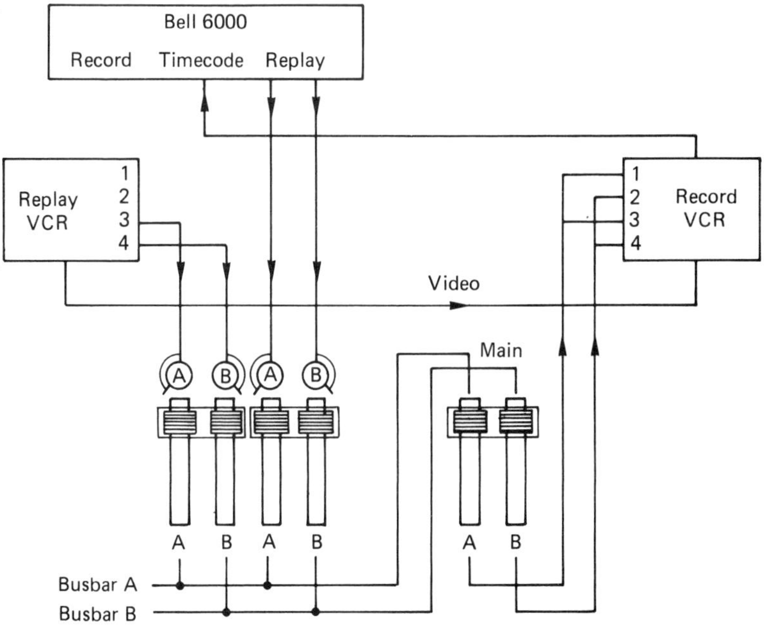


(b)

**Figure 9.17** Record timecode and stereo audio onto digital recorder during transfer (a). Timecode from record machine triggers digital replay within 30 μs (b). Stereo audio from the first section can be mixed with that from the second.

## *The future*

Although the Bel 6000 is a very elegant system, the real need is for the video editor only to use the sound as a guide, with the dubbing mixer relaying all the sound in the dubbing suite. This is very time-consuming and expensive but it is technically possible and it is getting cheaper.

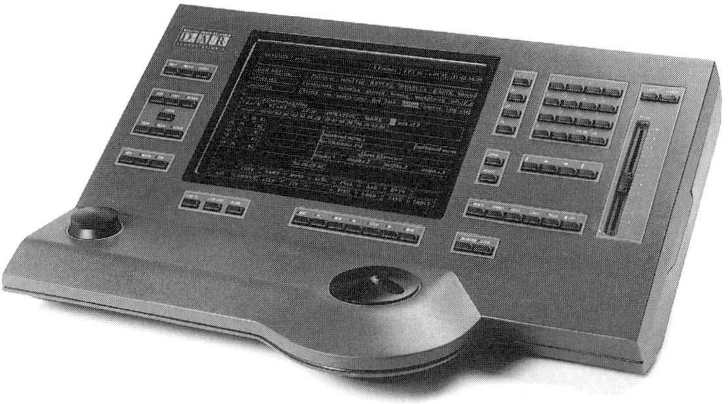

**Figure 9.18** DAR Soundstation. Solid state 8-track recorder/player.

One can foresee a time when digitally recorded field tapes are compiled directly onto digital recorders. There are video edit controllers that can store the timecode values for their edits on computer disc. Companies like Audio Kinetics are working on systems that will use this information to compile audio dub masters directly from the field tapes. The compiled master could be four tracks of a multitrack or Nagra Digital or a pair of DAT or timecode open reel machines, although the ultimate would be to use a hard-disc based digital device, like the DAR Soundstation. The software could allow for overlaps and the data could produce a dubbing list for the mixer.

# Glossary

**A-signal** Conventionally, the left signal of a stereo pair, or channel 1 of a dual mono pair.
**AB pair** A pair of signals that can be fed directly to a pair of spaced loudspeakers to create a stereo effect, or a pair of microphones that will generate such a signal.
**Acoustic** Conventionally, the subjective acoustical behaviour of a room that results from its reverberation characteristics, shape and size.
**AES/EBU Digital Interface** A process of converting an analogue stereo signal into digital data that can be sent serially over existing (mono) lines and connectors.
**Ambience** In stereo, the subjective impression of the acoustic environment within which the sound stage is placed. It can imply lack of specific direction for sound sources.
**Ambisonics** A system of recording and reproducing surround-sound information. Calrec have adopted this system for their soundfield microphone.
**Aspect Ratio** The ratio between the width and height of a television image (4:3 for terrestrial television).
**Audio Vectorscope** A device for displaying, usually as a trace on a CRT, the amplitude and phase characteristics of a stereo pair of signals.

**B-signal** Conventionally, the right signal of a stereo pair or channel 2 of a dual mono pair.
**Balanced Circuit** A circuit in which the signal is carried by a pair of conductors neither of which is connected to earth.

**Baseband** The range of audio frequencies that will be used to modulate a carrier.
**Bass Tip-up** An increase in the bass response of a directional microphone as the distance from a sound source decreases. It is a function of the pressure gradient employed to obtain directivity.
**Betacam SP (Superior Performance)** A cassette-based helical scan recording system that provides a pair of high quality FM audio channels in addition to Longitudinal Tracks (q.v.) with Dolby-C noise reduction.
**Binaural** A system for obtaining directional information by placing microphones to simulate the position of human ears, either by the use of a dummy head or flat disc of appropriate size. The signal obtained is designed to be listened to on headphones.
**Blumlein Technique** Named after Alan Blumlein, a technique for obtaining directional information by the use of a Crossed Pair (q.v.) of figure-of-eight microphones.

**Cannon** Another name for an XLR (q.v.)
**Cardioid** A Polar Response (q.v.) that gives a microphone the ability to pick up sound over an angle of 120°–150° and discriminates against sounds coming from behind.
**Chrominance Signal** A video signal containing information about the colour components of the total signal.
**Close Up (CU)** A shot that frames the head of an individual.
**Cocktail Party Effect** A Psycho-acoustic (q.v.) effect that will subconsciously direct the attention of the brain to a particular sound, and so brings its perception into the conscious domain.
**Coincident Pair** A pair of microphones positioned so that their diaphragms occupy effectively the same point in space with respect to a sound wave.
**Colour Difference Signals** Video signals that result from subtracting the luminance component from the red and blue components.
**Comb Filtering** An effect obtained by adding a pair of similar signals with a timing difference, resulting in notches in the frequency spectrum of the summed signal due to phase cancellation.
**Compander** A device that compresses the dynamic range of a signal prior to recording, and expands it back to its original value after playback. The technique is widely used in noise reduction devices.

**Component Analogue** A form of video signal processing in which the Chrominance (q.v.) components are kept separate from the Luminance (q.v.) component.
**Composite** A form of video signal processing in which the Chrominance (q.v.) components are superimposed (by some form of modulation) onto the Luminance (q.v.) component.
**Compression Ratio** The range of dB through which the input to a compressor has to vary to produce a 1 dB variation in output level.
**Compressor** A device that reduces the dynamic range of a signal.
**Control Track** A Longitudinal Track (q.v.) recorded along the edge of a tape by a helical scan recorder to permit the rotary heads to follow the diagonal recorded tracks on replay.
**Crossed Pair** A pair of Coincident (q.v.) microphones with polar responses designed to produce AB stereo signals.
**Crossfade** To alter the relative gains of two channels so that there is a gradual change from one signal to the other.
**Crosstalk** The breakthrough of one signal onto another. When the signals concerned are a stereo pair the result is to reduce channel separation.

**dB** Abbreviation for decibel: a measure of the relative strengths of two signals, expressed logarithmically.
**dBu** A decibel unit expressing the absolute voltage level by comparing it with 0.775 volts.
**Decca Tree** An arrangement of spaced omnidirectional microphones designed to provide a stereo signal and at the same time minimize the Hole-in-the-middle Effect (q.v.).
**De-emphasis** The process, complementary to pre-emphasis, which restores the original frequency response of the signal.
**Detachable Capsule Mic** A microphone with a capsule that can be detached and operated at a distance from the main body, to which it is connected by a thin, flexible lead.
**Deviation** The change in frequency of an FM carrier due to the amplitude of the modulating signal. The maximum deviation permitted of a system is determined at its design stage.
**Difference Signal** The signal that results from subtracting the B signal from the A signal of an AB pair. Also the signal generated by the sideways-facing element of an MS microphone pair.

**Direct Sound** Sound arriving at the ear directly from a source without experiencing any reflection.
**Dolby-C** A method of noise reduction using companders, employed in component recording systems. It can reduce noise by up to 20 dB.

**Echo** A reflected sound that arrives sufficiently delayed after the Direct Sound (q.v.) for it to be heard as a separate signal (cf. Reverberation).
**Edge Images** Sound images positioned to the sides of the reproduced sound stage.
**End-fire** A microphone that is pointed towards the sound source.

**Fishpole** An extensible pole, fitted with a swivel, enabling an operator to position a microphone close to the action while personally remaining out of shot.
**FM Channels** The means whereby two separate audio signals can be recorded onto tape as a frequency modulated signal, occupying a separate part of the frequency spectrum from the chrominance signal, but the same space physically on the tape.
**Framing** The process of composing images within the 4:3 aspect ratio to convey the production intent.
**Frequency Response** The variations in amplitude of a signal, from the original, with variation in frequency.

**Gang** To link controls together so that they move and act as one on a stereo signal. To electronically link dynamic gain control devices (e.g. compressors) so that gain variations in one channel are copied in the other.

**Haas Effect** The psycho-acoustic effect whereby the brain perceives a sound source as being in the direction of the first sound to arrive, even if a replica of that sound arrives louder but delayed (within limits).
**Handling Noise** Low frequency noise that results from handling microphones and fishpoles. It causes more problems in stereo than with mono because of the bass tip-up of the directional elements of the capsules.
**Helical Scan** A method of recording in which the tape is wrapped in a part spiral around a spinning drum, containing the

record/replay heads. The signal is recorded as a series of diagonal stripes.
**Hole-in-the-middle Effect** An effect that can be produced by spaced pair microphones whereby the stereo image is exaggerated, with centrally placed images appearing to retreat into the distance.
**Hypercardioid** A Polar Response (q.v.) that gives a microphone the ability to pick up sound over an angle of 90°–120°, discriminates against sounds far round to the sides, but which is sensitive to sounds coming from directly behind.

**Image Wander** Spatial instability of a stereo image brought about by differences in the frequency response of a stereo pair, or by poor microphone placement in a Multi-mic set-up (q.v.)
**Interaural Differences** Differences in amplitude and phase perceived at the ears that result from either interchannel differences of a stereo pair, or the offsetting to one side of a real sound source.
**Interchannel Differences** Differences in amplitude and phase of a stereo pair that will place the image in the stereo scene when interpreted by the brain.

**Limiter** A Compressor (q.v.) having a compression ratio greater than 10:1.
**Lobe** An individual component of a polar response.
**Long Shot (LS)** A shot that frames an entire standing individual.
**Longitudinal Speed** The speed at which tape moves through the transport system of a helical scan recorder.
**Longitudinal Track** A track (audio, timecode or control) that is recorded by stationary heads within a helical scan recorder.
**Luminance Signal** A video signal containing information about the brightness of the total signal.

**MII** A cassette-based helical scan recording system that can provide a pair of digital sound channels in addition to high quality FM audio channels and Longitudinal Tracks (q.v.) with noise reduction.
**M-signal** The forward-facing component of an MS system. The monophonic sum of A and B signals.
**Matrix** An electrical or mathematical process for converting

signals from one format to another, for example from AB to MS or vice versa.
**Metering** A means of providing information about a signal visually.
**Mic Axis** An imaginary line drawn from the capsule of a microphone through the most sensitive point in its polar diagram, or through the mid point of its AB or M response.
**Mid Shot (MS)** A shot that frames the head and torso of an individual down as far as the point of the breastbone.
**Monitoring** A means of providing information about a signal aurally (usually by means of headphones or loudspeakers).
**Mono Compatibility** The degree to which a stereo signal pair can be summed for mono without noticeable degradation. It depends on such factors as level and phase differences between the two channels of an AB signal, and on the direct to reverberant sound ratio.
**Mono Signal** A signal, or pair of signals, that contains no directional information.
**MS Pair** A pair of signals that represent the forward-facing and sideways-facing elements of a stereo signal. MS signals allow easier manipulation of image width, but cannot be fed directly to a pair of spaced loudspeakers to create a stereo effect, or a pair of microphones that will generate such signals.
**Multi-miking** A method of miking a scene using a number of microphones, positioned close to the individual sound sources.

**Near Coincident Pair** Microphones spaced apart by a distance equivalent to the spacing of the ears, but without the intervening dummy head of a binaural system.
**Near Field** The region of a sound field where direct sound predominates over reverberant sound.
**Noddies** Pick-up shots, usually of an interviewer simulating response to the person being interviewed, and provided for continuity, at the end of an interview to permit ease of editing.

**Off Mic** A source that is positioned within the non-sensitive part of a microphone's polar response is said to be off mic.
**Omni** A microphone that has a polar response substantially constant in all directions (vertically as well as horizontally).
**On Axis** A source that is positioned on, or close to, the Mic Axis (q.v.) is said to be on axis.

**Out-of-phase Region** That region of a stereo pair of microphones where a source of sound lies within lobes of opposite polarities.

**Pad** A network of resistors used to reduce the level of a signal.

**Pan Pot** A control on a mixer used to position a mono signal somewhere between fully-left and fully-right by varying the relative gains of the left and right channels.

**Peak Programme Meter** A meter having a short rise time and slow fallback time that provides an indication of peak signal levels.

**Personal Mic** A small microphone that can be worn inconspicuously or invisibly.

**Phantom Power** A system of powering microphones remotely, and which maintains both signal wires at the same potential so that dynamic microphones are unaffected if connected.

**Phase Meter** A meter giving an indication of whether a pair of AB signals are in or out of phase, or whether there is any phase correlation at all.

**Phase-reversed** If, at any point in the AB stereo chain, the polarity of one of the signals is reversed with respect to the other, the signal is said to be phase-reversed. The effect of phase-reversal is to make stereo image location impossible and to cancel centre images in mono to a large degree.

**Phasing** Another name for Comb Filtering (q.v.). Alternatively, the process of checking and ensuring that microphones are in phase.

**Pick-ups** Shots recorded after the main body of an interview or scene, usually with different camera angles to permit ease and variety of editing.

**Piggy-back Mount** A method of mounting microphones having different polar responses to provide a range of stereo acceptance angles.

**Polar Response** Graphical or descriptive representation of the sensitivity of a microphone to sounds coming from different angles.

**Post-fader Aux** A feed of a channel, or group of channels, taken off after the channel or group fader.

**Post-production** The processes of editing, copying and mixing the material shot on location to compile a complete programme.

**Pre-emphasis** A means of boosting the (usually) high frequency spectrum of a signal prior to recording in order to reduce (high frequency) noise (cf. De-emphasis).

**Pre-fader Aux** A feed of a channel, or group of channels, taken off before the channel or group fader.

**Pressure Gradient Microphone** A microphone whose output is proportional to the difference in pressure between the front and rear faces of the diaphragm.

**Psycho-acoustics** The relationship between sounds and their physiological and psychological effects.

**Radio Mic** A microphone, usually personal or hand-held, with an integral transmitter and remote receiver that permits freedom of movement for the wearer/user.

**Read Speed** The speed at which the rotating heads of a helical scan tape machine read the recorded information. It is much higher than the Longitudinal Speed (q.v.) of the tape.

**Reverberation** Reflected sounds that arrive so soon after the Direct Sound (q.v.), and in such numbers, that they are perceived to be an integral part of the sound (cf. Echo).

**Rifle Mic** A microphone with a highly directional polar response at mid and high frequencies, usually becoming cardioid at low frequencies.

**S-signal** Another name for the Difference Signal (q.v.).

**Sensitivity** Of a microphone, a measure of the output voltage obtained for a given sound level arriving on axis. It can also be used to describe the response of a microphone to sounds arriving other than on axis.

**Shotgun** Another name for Rifle Mic (q.v.).

**Side-fire** A microphone having its capsules facing sideways. They can usually be rotated with respect to each other to alter the width of the stereo image.

**Singles** Shots framed so as to cover one individual of a group.

**Sound Stage** The apparent width, in a stereo system, within which the physical action occurs.

**Spaced Pair** A pair of microphone spaced apart so that both time and level differences occur.

**Soundfield Microphone** A microphone having four capsules, arranged in a tetrahedron, that behave as coincident pairs, and whose polar responses and apparent angles can be varied electronically (see Ambisonics).

**Split MS** A way of separating an MS pair of microphones so as to introduce Time-of-arrival (q.v.) differences.
**Spot Microphone** A mono microphone placed close to a sound source and panned so that its output appears in the correct place in the sound stage.
**Stereo Effect** The combination of amplitude and phase differences in a pair of signals that results in the production of clearly identifiable sound sources within the sound stage.
**Stereo Image** The apparent position of a reproduced sound source within the sound stage.
**Stereo Pair** A pair of signals, lines, microphones carrying the left and right signals of stereo programme material.
**Sum Signal** The signal that results from adding the A and B signals of an AB pair. Also the signal generated by the forwards-facing element of an MS microphone pair.
**S-VHS** A development of the VHS format that provides higher luminance definition, less cross-colour interference, and a pair of high quality FM audio channels.
**Sync Sound** Sound recorded at the same time as the picture so that visual and aural signals match.

**Timecode** A digital code, recorded as an audio bandwidth signal on a longitudinal track, and/or as a part of the video signal, and used to achieve sound and picture synchronization in post-production.
**Time-of-arrival Difference** The difference in arrival time, at the ears, of a signal coming from off-centre.
**'T' Power** (Also known as A/B power) A method of powering a microphone remotely that does not require a third conductor. Dynamic microphones do not work if they are connected to a 'T' power supply.
**Two Shot** A shot framed to include two people either in whole or in part.

**Volume Unit Meter** A meter with rise and fall times of 300 msec to give a reading that endeavours to simulate perceived volume.

**Wide Shot (WS)** A shot framed to display a panoramic view of a scene.
**Width Control** A control found on some mixing desks that

varies the gain of the S signal to vary the width of the stereo image. If a mixing desk has enough spare faders and aux sends, a width control can be constructed.

**Wildtrack** Non-synchronous sound, usually recorded after the main shoot, but with the cast present, for use in post-production.

**Write Speed** The speed at which the rotating heads of a helical scan tape machine record information. It is much higher than the Longitudinal Speed (q.v.) of the tape.

**XLR** An audio connector that is moisture resistant and when used in the conventional manner, pin 1 – earth in a Balanced Circuit (q.v.), will provide a protected (screened) circuit by making the earth connection first.

# Index